Materials Selection for Design and Manufacturing

MATERIALS SELECTION FOR DESIGN AND MANUFACTURING

Theory and Practice

JOSEPH DATSKO
University of Michigan
Ann Arbor, Michigan

MARCEL DEKKER, INC.   NEW YORK • BASEL • HONG KONG

Library of Congress Cataloging-in-Publication Data

Datsko, Joseph
   Materials selection for design and manufacturing: theory and practice/Joseph Datsko.
     p. cm.
   Includes bibliographical references and index.
   ISBN 0-8247-9844-9 (hardcover : alk. paper)
   1. Strength of materials. 2. Metals. I. Title.
TA405.D326   1997
620.1'12—dc21

                                                  96-50430
                                                      CIP

The publisher offers discounts on this book when ordered in bulk quantities. For more information, write to Special Sales/Professional Marketing at the address below.

This book is printed on acid-free paper.

Copyright © 1997 by MARCEL DEKKER, INC. All Rights Reserved.

Neither this book nor any part may be reproduced or transmitted in any form or by any means, electronic or mechanical, including photocopying, microfilming, and recording, or by any information storage and retrieval system, without permission in writing from the publisher.

MARCEL DEKKER, INC.
270 Madison Avenue, New York, New York 10016
http://www.dekker.com

Current printing (last digit):
10 9 8 7 6 5 4 3 2 1

PRINTED IN THE UNITED STATES OF AMERICA

PREFACE

This book is written to present to engineers who are involved with the design or manufacture of mechanical devices some new concepts that will enable them to select materials on the basis of optimal fabricability as well as achieve optimal properties in the *fabricated* part. These concepts have been developed by the author during the past 40 years while teaching and supervising engineering students on individual research projects at the University of Michigan. These new concepts are based on a thorough knowledge of the relationships of material properties to the microstructure of the material.

      The introductory chapter discusses the important role that a knowledge of material properties and manufacturing processes plays in achieving the ideal design and fabrication of a mechanical product. This critical role is illustrated with several examples. Also included in this chapter are the design rules for optimal producibility.

      This book contains several modules of discrete subject matter. The first module is primarily a review of the basic concepts of material science that are needed to understand how the microstructures of a material are altered during processing. This module presents the subject matter in a series of application rules that assist the reader in becoming proficient in selecting or specifying the best metal for a given part. Conversely, these same rules aid in diagnosing the cause of a part failure or a manufacturing problem. These concepts are presented in Chapters 2 through 6.

iv   Preface

The second and third modules consist of the concepts that constitute a new engineering discipline that the author refers to as Strength Analysis—*the calculation of the strength of the material in the fabricated part*. The calculations are made on the basis of the original properties of the metal while in the shape of a bar or plate and the changes in the properties that result from thermal or deformational processing of the metal.

The second module presents the concepts of Strength Analysis following thermal processing. Chapter 6 introduces the techniques for calculating the strength of steel and Chapter 9 discusses the methods for predicting the strength and load capacity of weldments. Module one, or an equivalent knowledge of microstructures, is a prerequisite for the second module.

The third module presents the concepts of Strength Analysis following deformational processing (cold work). This module includes Chapter 7, Mechanical Properties, and Chapter 8, Deformation Strengthening and Formability. Chapter 7, which can stand alone, introduces several new relationships between the various individual mechanical properties. These new relationships form the basis for treating the subject of mechanical properties as a science. They make it possible for an engineer to examine published or experimental values of the mechanical properties of a metal and to determine if the values are reasonable or whether they are definitely incorrect.

Chapter 8 expands the subject of mechanical properties by introducing several new concepts concerning the sense (tensile or compressive) of the strength, the direction in which the strength lies (longitudinal or transverse), and the prior history of the deforming strains. This chapter presents an original code for the designation of strength that includes all of the above factors. It also presents the author's "Apparent Rules of Strain Strengthening." These rules make it possible for an engineer, for the first time, to predict with reasonably good accuracy the complete strength distribution in a cold formed part when the strain history is known.

Chapter 10, which constitutes the fourth module, includes a fundamental relationship of a metal's machinability to its primary physical properties. This relationship makes it possible for a design engineer to select the optimal metal for a component part, including the evaluation of the machinability of the metal. It also makes it possible for the designer to evaluate the reliability of the published values of the metal's machinability ratings. The relationships of the tool material, tool shape, and the size of cut to the metal's machinability make it possible for a manufacturing engineer to optimize the actual cutting conditions for any machining operation.

Some representative cost-of-materials data are included along with the mechanical property data. The integration of mechanical properties, fabricability, and cost is of prime importance in achieving the optimal design of mechanical components. The author refers to this interdisciplinary area as *Strengthonomy*, which is defined as the engineering science of selecting materials for mechanical components on the basis of achieving the optimal properties in the fabricated parts along with the least cost of fabrication and minimal weight.

In summary, this volume presents the analytical tools that enable engineers to specify the best metal from which structural and mechanical parts should be made. The use of these tools ensures the least cost or weight and the maximum reliability. Relating a metal's weldability, formability, and machinability to its physical properties makes it possible to optimize the specification of a material based on its composition.

Developed as a textbook developed for use in a one-semester engineering college course, this volume is also recommended as a resource book for practicing product design, manufacturing, materials, and quality control engineers.

I want to acknowledge the valuable contributions of my colleague Professor William Mitchell, who worked with me on much of the research reported in this textbook. I also want to acknowledge the valuable assistance of Ms. Charlotte Wahlstrom in preparing the manuscript for publication.

Finally, my public thanks to my family for their patience and understanding during the many busy years prior to the publication of this book.

Joseph Datsko

CONTENTS

Preface   *iii*

Chapter 1   **MATERIALS AND MANUFACTURING CONSIDERATIONS IN PRODUCT DESIGN**   1

    Introduction   1
    Application of the Science of Mechanical Properties to
        Improved Design for Manufacturability   3
        Selection of the Optimal Material   3
        Improving Mechanical Reliability   5
        Selecting Alternative Materials   6
        Reducing the Weight of Mechanical Components   6
        Computer-Aided Design (CAD) and Manufacturing (CAM)   7
    Datsko's Design Rules for Optimal Producibility   7
    References   23

Chapter 2   **THE STRUCTURE OF SOLIDS**   25

    Introduction   25
    Atomic Bonding Forces   25
        Ionic Bonds   26
        Covalent Bonds   28
        Metallic Bonds   30
        Molecular Bonds or van der Waals Bonds   32

viii   Contents

Atomic Structure	33
Amorphous Solids	34
Molecular Solids	34
Crystalline Solids	42
Crystal Imperfections	53
Slip in Crystalline Solids	59
Mechanical Strength	63
Principles of Mechanical Strength	63
References	67

Chapter 3   EQUILIBRIUM PHASE DIAGRAMS   69

Introduction	69
Phases	70
Heating and Cooling Curves	71
Isothermal Phase Change	71
Non-Isothermal Phase Change	75
Binary Phase Diagrams	76
Three Application Rules	78
Some Common Phase Diagrams	81
References	89

Chapter 4   MICROCONSTITUENT DIAGRAMS AND MICROSTRUCTURES   91

Introduction	91
Microstructures	92
Microscopy	92
Microconstituent Diagrams	96
The Lead-Tin Microconstituent Diagram	96
The Iron-Carbon Microconstituent Diagram	99
Rules of Mechanical Properties	101
Ferrous Equilibrium Microconstituents	103
Austenite	104
Ferrite	104
Cementite	104
Pearlite	104
References	111

Chapter 5 NON-EQUILIBRIUM DIAGRAMS AND MICROCONSTITUENTS — 113

Introduction	113
Diffusion and Transformation Rates	114
Non-Equilibrium Phases and Microconstituents	121
Medium and Fine Pearlite	121
Martensite	121
Bainite	124
Isothermal Transformation Diagrams	125
Hardenability	136
Tempering	141
Example Problems	150
References	152
Study Problems	153

Chapter 6 THERMAL STRENGTHENING AND COSTS — 155

Introduction	155
Ferrous Metals	155
Tempering	156
End-Quench Hardenability Limits	157
Hardenability and Composition	161
Cost of Steel	166
Non-Ferrous and Non-Metallic Materials	169
Example Problems	171
References	174
Study Problems	175

Chapter 7 MECHANICAL PROPERTIES — 177

Introduction	177
Hardness	178
Rockwell Hardness	178
Brinell Hardness	180
Meyer Hardness	181
Diamond Pyramid or Vickers Hardness	184
Knoop Hardness	185
Scleroscope Hardness	186
The Tensile Test	187

Engineering Stress-Strain	188
True Stress-Strain	191
Tensile Properties	199
Modulus of Elasticity	200
Proportional Limit	200
Elastic Limit	200
Yield Strength	200
Yield Point	201
Tensile Strength	202
Fracture Strength	202
Reduction of Area	202
Fracture Strain	203
Percent Elongation	203
Strain-Strengthening Exponent	204
Stress Coefficient	204
Strength-Stress-Strain Relationships	204
Nominal Strain and Area	205
Natural Strain and Nominal Strain	206
True Stress and Nominal Stress	207
Strain-Strengthening Exponent and Maximum Load Strain	207
Tensile Strength, Stress Coefficient, and Strain-Strengthening Exponent	209
Yield Strength and Percent Cold Work	209
Tensile Strength and Percent Cold Work	211
Tensile Strength-to-Brinell Hardness Ratio	214
The Effect of Plastic Deformation on σ_o and m	218
Predicting σ_o and m After Tensile Deformation	222
Fatigue Properties of Metals	225
Fatigue Strength	226
Endurance Limit	226
Fatigue Ratio	226
Intrinsic Fatigue Strength	228
Impact Strength	230
Creep Strength	231
Estimating Low Cycle Fatigue Strength from Tensile Test Data	235
The Current Method	235
Datsko's Method	235
Example Problems	240

		Contents	xi
	References		243
	Study Problems		243

Chapter 8   DEFORMATION STRENGTHENING AND FORMABILITY   247

Introduction	247
Datsko's Strength Designation	250
Datsko's Apparent Rules of Strain Strengthening	253
Example of Strength Analysis	259
Determination of Deformation Strains	260
Axial Deformation	261
Rolling Deformation	261
Bending Deformation	261
Torsion and Shear Deformation	267
Forces, Work, Energy During Deformation	272
Example Problems	275
References	279
Study Problems	280

Chapter 9   STRENGTH OF WELDMENTS   287

Introduction	287
The Welding Processes	288
Types of Welding Processes	288
Calculation of the Size of the Weld	292
Burn-Off Rate (BOR)	293
Bead Size	293
Depth of Penetration	295
Axial Load Capacity of Butt Joints	295
Strength and Microstructures	297
Fillet Welds	300
Fillet Size	300
Load Capacity of Fillet Welds	301
Example Problems	302
References	304
Study Problems	304

xii   Contents

Chapter 10 MACHINABILITY OF METALS 307

    Introduction 307
    Definition of Machining 308
    Cutting Conditions 308
    Chip-Forming Machine Tools 310
    Tool Life 311
    Tool Shape and Cutting Efficiency 316
        Bake-Rake Angle 317
        Relief Angles 317
        End-Cutting-Edge Angle 317
        Side-Rake Angle 318
        Side-Cutting-Edge Angle 319
        Nose Radius 320
    Machinability 320
    A Fundamental Machinability Equation 324
    Example Problems 333
    References 335
    Study Problems 336

APPENDIX A Conversion Factors: U.S. Common to S.I. Units 339
APPENDIX B B-1 Physical Properties of Metals 340
                B-2 Tensile Properties of Some Metals 341
                B-3 Tensile Properties at Low Strain Rates and
                      Elevated Temperatures 344
                B-4 Mechanical Properties of Some Plastics 345
APPENDIX C C-1 Conversion from Cold Work to Strain 346
                C-2 Percent Cold Work (CW) vs. Natural Strain (ε) 347
APPENDIX D D-1 Designations for Aluminum Alloy 348
                D-2 Chemical Composition of Some Aluminum
                      Alloys 349
                D-3 Typical Mechanical Properties of Wrought
                      Aluminum Alloy 349
APPENDIX E E-1 AISI and SAE Designation of Structural Steels 350
                E-2 Composition of Some AISI Steels 351
APPENDIX F F-1 Approximate Prices of Castings 352
                F-2 Warehouse Prices of Steel 352
                F-3 Market Prices of Bulk Plastics 352
APPENDIX G Forming Operations 353
APPENDIX H Machining Operations 359

Index *367*

Chapter 1

MATERIALS AND MANUFACTURING CONSIDERATIONS IN PRODUCT DESIGN

INTRODUCTION

Historically product designers have not been expected to specify the manufacturing processes to be used to fabricate the parts they design. Management's philosophy has been based on the concepts that (1) the design engineer does not know enough about the manufacturing processes to specify them, and (2) the company already employs production engineers who are specialists in the details of the manufacturing processes. But the fact remains that product design engineers determine, sometimes unknowingly, the processes by which the parts will be manufactured. Lack of attention at this important stage can result in overlooking or not employing the most economical processes.

In an attempt to alleviate this problem, many of the large manufacturing companies have formed producibility groups whose function is to evaluate the preliminary designs of the engineering department before they are released to the manufacturing department. By this means components that are difficult, if not impossible, to fabricate are avoided. However, evaluations of the preliminary designs of the traditional product designer by a producibility group before finalizing the designs seldom results in as optimal a design as when the design engineer is thoroughly educated in the engineering fundamentals of materials and manufacturing processes.

when the design engineer is thoroughly educated in the engineering fundamentals of materials and manufacturing processes.

Open competition increasingly requires that product design engineers, as they create and detail the design, knowingly select the processes by which parts will be manufactured. And, intentionally, product design engineeers will be selecting the most economical processes and materials. To do this, engineers need to be equipped with a knowledge of the science of mechanical properties.

If engineers are to design mechanical components that are functional and reliable, they must be knowledgeable in the following areas: stress analysis, kinematics, kinetics, friction and wear, corrosion, materials properties, and strength analysis. All of these areas, with the exception of strength analysis, are taught in most engineering colleges. Strength analysis, the calculation of the strength of the material in a fabricated part, is an engineering discipline now being developed.

If engineers are to be proficient in designing mechanical components that are also economically producible, they must be knowledgeable in the following additional areas: manufacturing processes, relationships between material properties and fabricability, relationships between operating variables and fabrication time, cost of materials, material properties, and strength analysis.

Until now, the subject matter of these latter six areas has generally been taught and used either *qualitatively* or else as unrelated bits of factual data. However, many science-based *quantitative* relationships in these six areas have been developed by the author during the past 30 years. These concepts and relationships form the beginning of a science base to the subject of mechanical properties, which in this context includes both the primary properties associated with strength — ductility and hardness — and the secondary properties (or characteristics) of formability, weldability, and machinability. Although these properties apply to various classes of materials, the focus in this book is on metals.

This science of mechanical properties now, for the first time, enables the engineer to perform a large number of important quantitative analyses over a broad spectrum of design functions that directly affect the reliability, cost, weight, and material utilization of component parts. These are important ingredients in designing for manufacturability. The following section illustrates how this new science can greatly improve the quality of mechanical designs.

Materials and Manufacturing Considerations in Product Design

APPLICATION OF THE SCIENCE OF MECHANICAL PROPERTIES TO IMPROVED DESIGN FOR MANUFACTURABILITY

Selection of the Optimal Material

The design engineer often has difficulty in selecting the optimal material for mechanical components because much of the mechanical property data for materials is incomplete or incorrect. This is true for both the primary properties of strength and the secondary properties of machinability and formability. The engineer using unreliable data of this type will create mechanical components that are unreliable as well as uneconomical to manufacture. Examples of incorrect or unreliable mechanical property data published in the technical engineering handbooks and pamphlets are presented here to illustrate this problem.

Example 1. A certain company manufactures and sells a cobalt base superalloy. Its engineering catalogue lists the following tensile properties for the metal in the "as drawn" condition:

Cold work (percentage)	0	10	20	30	50
Yield strength (ksi)	100	109	120	130	170
Tensile strength (ksi)	145	170	190	215	250

The science of mechanical properties demonstrates that such a combination of properties is impossible. The author verified this by procuring and testing some of this annealed metal. The experimentally determined 0.2% offset yield strength was found to be 48 ksi and the ultimate tensile strength was 112 ksi. The published value of yield strength is more than 100% greater than that found experimentally in the author's laboratory. This is indeed a large discrepancy — too large to be neglected by a design engineer.

Example 2. The first example illustrates unreliable data in a material manufacturer's engineering pamphlet. This one illustrates incorrect mechanical property data in a research paper on the true stress-strain properties of several commercial titanium and maraging steel alloys that was published in the transactions of a professional engineering society. The data includes the yield strength and also the plastic strength coefficient σ_o and strain strengthening exponent m in the stress-strain equation $\sigma = \sigma_o \varepsilon^m$

The properties are given for 41 conditions of the titanium and maraging steel alloys.

It can be proven by means of the science of mechanical properties that all of the 41 strength coefficient values are incorrect and impossible, and the error is of the order of magnitude of 100%. Unfortunately, the correct values cannot be determined from the data given in the research paper. It should be noted that "no data are better than incorrect data for mechanical design calculations.

***Example* 3.** This example illustrates conflicting information regarding the machinability of metals. A United States Government agency sponsored two projects at two different airplane manufacturing companies to study the machinability of a nickel-base alloy (Rene 41) and a cobalt-base alloy (HS 25). Both studies were based on the cutting speed-tool life criterion of machinability. For the same tool material, one study found that the nickel alloy was 50% better, while the other reported that the cobalt alloy was 50% better. Now, which is more machinable? The nickel alloy or the cobalt alloy?

Consider the confusion surrounding the machinability of such a common metal as plain, low-carbon steel. Is it more machinable in the annealed (hot-rolled) condition or in the cold-drawn condition?

One metals handbook states that low-carbon, cold-drawn steels are significantly more machinable than hot-rolled bars. But another manual on cutting of metals has dozens of tables that show in each case, for all tool shapes, that the cold-drawn steel is significantly less machinable than the same hot-rolled steel.

Some steel manufacturers' handbooks give a higher machinability rating to cold-drawn steel. Others' show cold-drawn steels to be considerably less machinable. Which is correct? These examples demonstrate the problem confronting the design engineer attempting to select the optimal material for a given part on the basis of the material's machinability. Similar confusion exists in the field of forming.

To summarize, the science of mechanical properties enables the design engineer to: (1) determine whether the published mechanical property data for any material is reliable and valid, (2) more reliably assess the machinability and formability of a material on the basis of its primary properties rather than on misleading sales-motivated engineering data, and (3) select the material on the basis of the as-fabricated mechanical properties rather than on questionable published values.

Improving Mechanical Reliability

The reliability of mechanical components can be improved when parts are designed on the basis of the strength of the material in the fabricated parts instead of on the values listed in materials handbooks. Included in the science of mechanical properties is a body of analytical and empirical relationships that make it possible for a knowledgeable engineer to calculate the strength in any direction at any location in a fabricated part. Designs made from these calculations will be more reliable and will have more reliable factors of safety.

When a part is made by any of the cold-forming operations, the mechanical properties are greatly altered from their original values. The amount of cold work done in a forming operation is expressed as the percentage reduction of area or the percentage cold work. For any given material, the material manufacturer usually provides the design engineer with tabular or graphical data for the strength (yield and ultimate tensile) as a function of the amount of cold work. For example, a certain annealed metal has yield and tensile strengths of 35 and 80 ksi, respectively. The tabular data may show that these values are increased to 70 and 110 ksi with 20% cold work.

Suppose that a design engineer is planning on specifying this metal for a shaft that will have a center section cold-upset by 20% prior to milling gear teeth or splines on the enlarged section. What value should he use for the compressive yield strength in a circumferential direction at the root of the teeth? 70 ksi? And what should he use for the value of the tensile yield strength in the circumferential direction? The value of 70 ksi for the yield strength listed in the table is probably the value for the tensile yield strength in the longitudinal direction, although it is not clearly stated. Are the transverse properties the same as the longitudinal ones? And are the tensile and compressive values equal? They are not! They may be different by as little as 10% or as much as 70%, depending upon the particular metal.

In order for the engineer to design reliable mechanical components, he or she must know the answers to these questions and must be able to accurately calculate the strength in all directions in any formed part. The engineer must also be able to specify the actual fabricating process used to make the part because only by this means can it be assured that the part will receive the correct amount of cold work that has been designed into the part.

The calculation of the stresses in a loaded part is referred to as *stress analysis*. The calculation of the strength in a fabricated part is called *strength analysis*. The author has developed over the past 30 years a considerable number of analytical and empirical relationships that permit an engineer to make reliable calculations of the strength of the material after fabrication by many of the manufacturing processes, including welding and forming. The author has developed the *apparent rules of strain strengthening* by means of which an engineer can quite accurately calculate the value for any strength in any direction in a cold-formed part. He has also developed a method to calculate the load-carrying capacity of arc-welded members based on the welding conditions of arc voltage, current, and welding speed. By the application of these techniques, the reliability of mechanical components is greatly increased.

Selecting Alternative Materials

In the event of a shortage of one or more types of material, it is necessary to select an equivalent, non-strategic material. This selection can best be made on the basis of the concepts embodied in the science of mechanical properties, where the true material properties are presented in a coherent, analytically derived body of knowledge and where the deficiencies of the pseudo-properties, such as percentage elongation and hardness, are apparent.

Reducing the Weight of Mechanical Components

It is becoming increasingly more important to design and manufacture mechanical goods on the basis of minimal weight and avoidance of strategic materials. The concepts of strength analysis enable the engineer to be much more effective in achieving this design objective. In the application of cold-formed parts, which are widely used in the automotive and aerospace fields, a design is considered to be complete and acceptable when the most highly stressed regions are analyzed to ensure that the material specified is at a thickness sufficient to make the part safe. However, very little consideration is given to the likelihood that most of the material in the part has a low level of stress and is much overdesigned. A design based on strength analysis and the science of the manufacturing processes will have parts that utilize the most

Materials and Manufacturing Considerations in Product Design

economical materials; that are fabricated in such a manner that they have the maximal, realistically obtainable strength; and that have minimal thickness throughout so that the most favorable ratio of strength to stress is present at all locations, not just the so-called high-stress regions.

Computer-Aided Design (CAD) and Manufacturing (CAM)

The analytical and empirical quantitative relationships in the science of mechanical properties are ideally suited for computer analysis in the selection of optimal materials for component parts. Computer programs can be prepared to assist the design engineer in the selection of materials for families of parts such as stampings, extrusions, rolled sections, machined parts, and shafts. Programs can assist the manufacturing engineer in selecting the optimal processing conditions such as cutting speed, feed and tool shape in machining; welding speed, voltage, and current in arc welding; and die radius and amount of deformation in forming operations. Computer-aided design and manufacturing can significantly increase the productivity of discrete parts.

DATSKO'S DESIGN RULES FOR OPTIMAL PRODUCIBILITY

In addition to designing a mechanical device to reliably perform its function, the engineer should design for ease of producibility and minimal cost or weight. In order to achieve all of these objectives, the design must be an iterative process encompassing the following 11 design rules. In no case should the functional performance of the device be sacrificed in order to make it cheaper.

The design rules are first listed in outline form and then each is discussed in sufficient detail so that the reader can understand the significance of each.

1. Select the material on the basis of ease of fabricability as well as function and original cost.
2. Use the simplest configuration and specify standard sizes whenever possible.
3. Use configurations requiring the least number of separate operations.
4. Use configurations that are attainable with efficient manufacturing processes.

5. Design the fabrication processes to achieve the desired strength distribution in the finished part.
6. Provide clamping, locating, and measuring surfaces.
7. Specify tolerances and surface finish with consideration of the functional requirements as well as the manufacturing processes that must be used. Avoid tolerances less than ±0.002" whenever possible.
8. Determine the specific function of the part.
9. Determine the specific characteristics or features of a part that enable it to perform its function.
10. Prepare a process sheet listing in sequence the manufacturing operations needed to fabricate the part.
11. Evaluate the preliminary design by considering changes in the configuration that will simplify the fabrication. Then go through the design rules again. Repeat this process until the optimal design is achieved.

Rule 1. Select the Material on the Basis of Ease of Fabricability as Well as Function and Original Cost

The first consideration in the selection of the material is the functional requirements. If certain thermal, magnetic, corrosive, wear, and strength properties are necessary, then one must choose a material having the appropriate physical properties. Many materials may satisfy the functional requirements of the part, especially when strength is the dominating functional property. The final selection of the material from among all the candidate materials should be on the basis of the lowest total cost. In addition to considering the original base price of the material, one must consider the fabricability (secondary mechanical properties) of the material. For parts requiring extensive machining, the machinability of the material is very important. For example, a certain nickel-copper alloy (monel) and a copper-zinc alloy (brass) have similar corrosive, magnetic, and strength properties. However, the machinability of the brass may be 6 to 8 times as good as the monel.

If the part is fabricated by forming from a sheet or plate, then the formability of the material should be considered. When cold-forming processes are used, the design engineer should include in the stress analysis the increased strength that results from the ulastic (plastic deformation below the recrystallization temperature) deformation. Likewise, the weldability or the

Materials and Manufacturing Considerations in Product Design

castability of the material is important if the part is to be fabricated by welding or casting.

Rule 2. Use the Simplest Configuration and Specify Standard Sizes Whenever Possible

The overall design should be kept as simple as possible. Consideration should be given during the design stage to ensure that the mechanisms can be easily assembled and maintained. Avoid specifying features on the parts that are not really necessary. For example, specifying a fixed radius or chamfer on the end of a shaft simply to avoid a sharp corner may require a special tool and an additional operation to machine the shaft. On the design, simply adding an arrow pointing to the edge with the note "break corner" will enable the manufacturing department to select the least expensive method of rounding the end of shaft.

If cold-drawn or cold-rolled material can be used to avoid generating machined surfaces, select standard sizes and tolerances. For example, the commercial tolerance on a 1" diameter cold-drawn steel bar is +0.000, –0.002. Therefore, a dimension of 1.000" +0.000, –0.002 is preferable to the commonly specified dimension of 1.000" ±0.001.

Rule 3. Use Configurations Requiring the Least Number of Separate Operations

This rule is very important for both small-quantity and large-quantity production. In the case of very small quantity lot sizes, the actual set-up time may be greater than the fabrication or machining time. The following allowances for set-up times and other "non-productive" operations can serve as a guide to estimate the machining cost. This cost is important because machining is the most expensive of the manufacturing processes. For moderately simple parts, a skilled machinist should be allowed about a half hour after first receiving a blueprint to determine what machine tools will have to be used and to plan the sequence of machining operations. In addition to this, approximately a half hour is needed by the machinist to install the proper work-holding devices and cutting tool holders on each machine tool for each group of separate machining operations. And finally, measuring and inspecting time of about one minute must be allowed for each operation in addition to the actual

10   Chapter 1

chip-cutting time. Thus it is not uncommon for a machinist to spend six hours making one part on which only 15 minutes of actual chip-cutting time is required.

In the case of very large quantity production, the machine set-up time is negligible. In fact, in most cases special-purpose machine tools are used which are especially designed to machine only one specific part. The cost of the material is the major cost of producing a part in very large quantities. The handling time and the actual machining time are the other significant cost factors. Cost reduction in these three areas is most beneficial.

Rule 4. Use Configurations that Are Attainable with Efficient Manufacturing Processes

By knowing the relationship of the physical properties of the material to its fabricability and the "mechanisms" of each of the manufacturing processes, the design engineer can specify the optimal part configuration that will both fulfill the functional requirements and minimize the manufacturing cost. Small changes in the configuration of a part frequently can permit the use of a more efficient manufacturing process. The design function must be an iterative process of first considering one configuration along with the processes necessary to fabricate it and then considering a more efficient process and altering the configuration to enable that new process to be used.

Figure 1-1a illustrates a design of a bracket that in most cases would require considerable machining from bar stock, a casting, or a forging — or if it were made of a soft metal, by machining from an extrusion. Figure 1-1b illustrates how a small change in the configuration of the part makes it possible to more economically fabricate the bracket by simple bending operations on flat bar stock. The note included with the drawing in Figure 1-1b is necessary since during bending the width of the bar decreases slightly at the outer fiber and increases at the inner fiber.

Rule 5. Design the Fabrication Processes to Achieve the Desired Strength Distribution in the Finished Part

The welding processes and the forming processes alter the strength of the material so that the properties in the fabricated part are not the same as those in the original material. All the fusion welding processes create a heat-affected

Materials and Manufacturing Considerations in Product Design 11

zone in the weldment. The material in the heat-affected zone may be considerably stronger or weaker than the original base metal, depending upon the composition and condition of the metal and the cooling rate after welding. A proper design of a weld joint would include specifications of the size of the bead or fillet and the depth of penetration so that these values, in conjunction with the microstructures present, will give the weld joint the required load-carrying capacity.

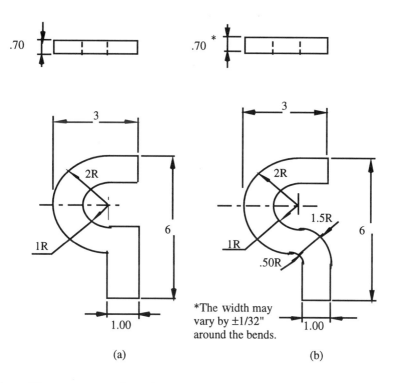

Figure 1-1. Design of a bracket. (1) Original Design. (b) Redesigned to be made by bending.

Even greater opportunities for improved design by using the concepts of strength analysis are present when parts that are formed from sheets and bars are specified. The mechanical properties of a material are greatly altered during room temperature deformation. For example, the yield strength may be increased by as much as 1,000% for some of the common metals when they are

12   Chapter 1

subjected to 50 or 60% deformation. In most of the forming operations, the deformation is not uniform throughout the fabricated part, and therefore the strength also varies throughout the part. The situation is much more complex because the tensile and compressive strengths are not equal and the properties in the longitudinal direction are different from those in the transverse direction. Therefore, the optimal design of a formed part would be one in which the engineer selects and specifies all the details of the fabrication of the part so that it has the maximum strength possible at the locations where the stresses are large.

Rule 6. Provide Clamping, Locating, and Measuring Surfaces

This rule is sometimes overlooked by the traditional product designer, with the result that the manufacturing department has to expend greater effort to fabricate the part. A classic example of how a design of a part that does not have locating or measuring surfaces can adversely affect the function of the part is the standard constant stress fatigue specimen illustrated in Figure 1-2.

In order for a constant stress fatigue specimen to be reliable (that is, to provide an accurate value of the stress that causes failure for a given number of cycles of load), it is necessary for the extrapolation of the two edges of the gage section to intersect at the centerline of the applied load. If the extrapolated edges intersect beyond the center of the load then the stress will not be constant over the entire gage section but instead will be at a maximum near the radii at the end nearest the load. And since the actual stress is not equal to the calculated stress, the resulting experimental data will not be reliable.

To better understand what locating and measuring surfaces are, consider a blueprint of a flat tensile specimen that has the following dimensions at the gage section: length, 2.000 ±0.100"; width, 0.750 ±0.002"; thickness, 0.125 ±0.001. These dimensions can be very easily measured by both the machinist who fabricates the specimen and by the engineer or technician who conducts the tensile test. The sides of the gage section are measuring surfaces — the anvils of a simple micrometer can be placed on them and the width can be accurately measured to a fraction of a thousandth of an inch.

Now consider the fatigue specimen shown in Figure 1-2. How will the engineer or technician who is going to use the specimen determine if the specimen is correct? How will he or she measure or inspect it to determine where the tapered sides intersect? The author's personal observation is that the person who uses the specimen simply assumes that it is machined correctly because he does not see any apparent way to "measure" it. That is, he does not

Materials and Manufacturing Considerations in Product Design 13

observe any surfaces to place the measuring instruments against to determine the value of the linear dimensions. Each tapered section is defined by two points that are 2-1/4" apart. However, the two points are only "imaginary" — one cannot use a micrometer or dial indicator to determine whether the one point that is dimensioned 3/8" is exactly 0.375" or whether it is in error by ±0.030".

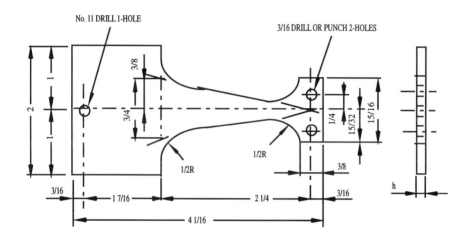

Figure 1-2. A standard constant stress flat fatigue specimen.

Consider also the plight of the machinist who has to make the specimen. Again, it has been the author's personal observation that the typical skilled machinist produces the gage section in the following manner. He first paints the "blank" specimen with a very rapidly drying paint called "layout blue." He then goes through the exact procedure with a "scriber" and scale to draw the gage section on the blank that a draftsman does in making the original drawing. The machinist then clamps the blank in a vise on a milling machine with a very light force. He moves the blank in the vise until the scribed line representing the gage section is parallel to the top of the vise jaws. He does this visually. He then proceeds to advance the milling cutter through the blank until the machined surface appears to coincide with the scribed line. The same procedure is followed for the second side. Consequently, neither the maker nor the user of the specimen actually "measures" the gage section. In summary, the specimen shown in Figure 1-2 is designed as a good project for a draftsman to draw but a poor project for a machinist to fabricate.

14   Chapter 1

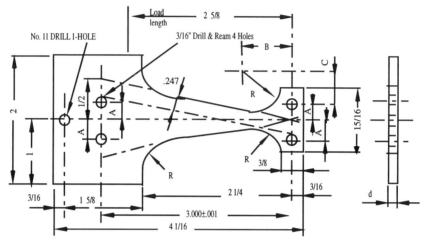

R=1/2", A=0.250 ± 0.001", B=0.6666", C=0.3626"
Note: All decimal dimensions are ± 0.001"

Figure 1-3. Redesigned constant stress fatigue specimen.

Figure 1-3 shows one way to "design" a constant stress fatigue specimen so that the accuracy of the gage section can easily be measured to a fraction of a thousandth of an inch. To make this specimen, the machinist first drills the No. 11 hole and then drills and reams the four 3/16" diameter holes to a location accuracy of ±.001". To generate the tapered section, the machinist inserts a 3/16" diameter dowel pin in the top left reamed hole and one in the bottom right reamed hole. He then places the assembly between the jaws of a vise on a milling machine so that the 2 pins rest on the top surface of the vise jaws. To obtain the required 0.247" distance that is required to have the tapered surface intersect the center line of the load application line, the machinist feeds the cutter through the blank until it is 0.341" (0.247 + 0.1875/2) above the vise jaws. The opposite side is machined in similar fashion. The critical dimensions can be measured accurately to 0.001".

Having received the specimen according to this design, the engineer or technician can accurately measure the gage section by mounting the specimen on pins which are then supported on parallels in a manner very similar to that used by the machinist. Not only are the specimens more easily inspected or measured when made by this modified design, but they can be machined more rapidly and at less expense.

Materials and Manufacturing Considerations in Product Design

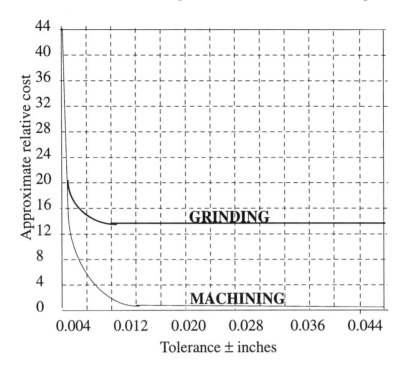

Figure 1-4. Chart showing the relative cost to produce parts as a function of the tolerance specified.

Rule 7. Specify Tolerances and Surface Finish with Consideration of the Functional Requirements as Well as the Manufacturing Processes that Must Be Used. Avoid Tolerances Less Than ±0.002" Whenever Possible

Figure 1-4 illustrates the approximate relative cost of machining a part as a function of the tolerance that a design engineer specifies on the drawing of the part. As a general practice, tolerances of less than ±0.002" should be avoided because of the significantly increased cost incurred in achieving such low tolerances. However, in the case of very high production, where the parts are machined on dedicated machine tools that are specifically built for the mass production of those parts, tolerances of ±0.0005" can easily be obtained. On the other hand, allowing tolerances greater than ±0.010" has a negligible effect on

Chapter 1

the cost and may lead to carelessness by machinists, who might not take the large tolerance seriously.

Table 1-1. Practical Tolerances and Surface Finishes Obtained by the Common Manufacturing Processes

	Process	Surface finish (rms)	Tolerance Good ± in.	Best obtainable ± in.
1.	Flame cutting	250-2000	0.060	0.020
2.	Sawing	126-1000	0.020	0.005
3.	Shaping	63-1000	0.010	0.001
4.	Broaching	32-260	0.005	0.0005
5.	Milling	32-500	0.005	0.001
6.	Turning	32-500	0.005	0.001
7.	Drilling	63-500	(0.010-000)L	(0.002-000)S
	(loc.)		0.015	0.002 (with jig)
8.	Reaming (dia.)	16-125	0.002	0.0005
9.	Hobbing	32--250	0.005	0.001
10.	Grinding	8-125	0.001	0.0002
11.	Lapping	0-16	0.0002	0.000,050
12.	Forming (brake)	Same as rolling	0.060	0.015
	(roll)	Same as rolling	0.010	0.005
13.	Stamping	Same as rolling	0.010	0.001
14.	Drawing	Same as rolling	0.010	0.002
15.	Forging	125-1000	0.060(in./in.)	0.030(in./in.)
16.	Rolling (cold)	8-32	0.010	0.001
17.	Extrusion (hot)	63- 250	0.020	0.005
	(cold)	8-63	0.005	0.001
18.	Sand casting	250-2000	1/8 in./ft.	1/32 in./ft.
19.	Permanent mold	40-125	0.03+ .002D	(0.010 + .002D)
20.	Die Casting	40-100	0.010	0.002 in./in.
21.	Investment casting	60-125	0.010	0.002 in./in.
22.	Sintered metal	40-100	0.005	0.001
23.	Sintered ceramic	30 250	0.030	0.020 in./in.
24.	Fusion welding	100-250	1/8	0.010
25.	Spot welding	Same as original	1/16	0.010
26.	Heat treating	Same as original	0.030	0.010

Materials and Manufacturing Considerations in Product Design 17

Grinding operations can produce parts with greater accuracy. Tolerances of ±0.0005" are easily obtained and tolerances greater than ±0.001" have a negligible effect on the cost. However, grinding is considerably more expensive than traditional machining for tolerances of more than approximately 0.001".

Table 1-1 lists the practical tolerances and surface finishes that are obtainable by the common manufacturing processes. The range of values for the surface finish are dependent upon the operating conditions such as feed and tool shape. The "good" column indicates the value of tolerance that is both reasonably small and fairly easy to obtain in most cases. The "best obtainable" column indicates the minimal values that can be obtained with care and without the need for additional sanding or filing.

Rule 8. Determine the Specific Function of the Part

Generally the design engineer finds this very easy to do. But often this step is neglected because of its simplicity, and as a result the fabricated part may not perform in the manner it was intended.

The following examples illustrate the sort of basic understanding needed about the function of parts. The function of a bolt is to transmit an axial tensile load in order to hold two pieces together. The function of a tensile specimen is to determine the tensile properties of the material it is made of. The function of the drive shaft in an automobile is to transmit torque from the transmission to the differential.

Rule 9. Determine the Specific Characteristics or Features of a Part that Enable It to Perform Its Function

This is a necessary activity if the functional reliability of the part is to be assured.

What are the characteristics or features of a tensile specimen that enable it to perform its function? The dimensions of the actual "gage" or "test" section must be accurately known. It is not enough to assume that the part has the exact dimensions that are given on the drawing or blueprint. The dimensions must be verified by the technician conducting the tensile test because the machinist may have made an error in machining the specimen. (The author has experienced this type of error. In machining a 0.357" diameter tensile specimen the machinist misread the reading on his micrometer by one revolution of the dial, an amount equal to 0.025". Consequently the actual diameter was only 0.332". The technician at a commercial testing laboratory assumed it was 0.357", as the drawing specified, and so he calculated the strength on the basis of the incorrect cross-sectional area.) Also, the area in the gage section must be uniform and the surface must be smooth and free of scratches or other defects.

The tensile specimen was an easy part to evaluate. The discussion of Rule 6, concerning the flat constant stress fatigue specimen, illustrates a more complex situation. The function of the fatigue specimen is to determine the fatigue strength of the material it is made of. In order to obtain reliable data from the test, the technician conducting the test must actually measure the thickness of the specimen. But more importantly, he or she must make certain that the two tapered surfaces extrapolate to intersect at the point where the load is applied. As discussed previously, the original design is very poor because it lacks the necessary measuring surfaces. The following example illustrates this problem.

Several years ago the author supervised a graduate student on a project to evaluate the fatigue strength of a cold-formed channel (U-shaped) section made of 1100-0 aluminum. The student took 25 rectangular blanks (4 1/16 × 2" × 1/8") and a drawing of the standard flat fatigue specimen to a machinist in a commercial machine shop and instructed him to make the specimens as shown in the drawing. After receiving the machined specimens, the student tested them and was dismayed to find that they all broke near the large end of the tapered section. Obviously the specimens were not machined so that the two surfaces of the tapered section intersected at the point of the load application. The purpose of a constant stress specimen is to show random failures along the length of the tapered section.

After this experience, additional specimens were machined according to the redesigned drawing, and in this case the failures were randomly distributed along the length of the gage section. Prior to testing each specimen, the student was able to determine to an accuracy of ± 0.001" the deviation of each from the desired dimensions.

Rule 10. Prepare a Process Sheet Listing in Sequence the Manufacturing Operations Needed to Fabricate the Part

The design of a mechanical component should be an iterative process. After creating the first design (drawing or sketch), the engineer should consider the

Materials and Manufacturing Considerations in Product Design   19

manufacturing processes that will be used to create the finished part. The following example illustrates this process.

Several years ago the author had a research project to study the springback that occurs after the plastic bending of metal. To accomplish this experimentally, it was necessary to make a clamp that could be attached to the 2" diameter column of a tensile testing machine and also have it hold the 3/8" diameter arm of a dial gage. The setup is illustrated in Figure 1-5.

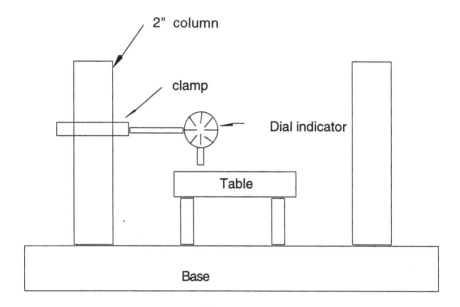

Figure 1-5. Sketch of a tensile testing machine with a clamp to hold a dial indicator.

Included in the group working on this project was a good (A average) graduate student who also had extensive machine shop experience. He was asked to design and fabricate the required clamp. The product that he created is shown in Figure 1-6. The following operations were required to make this clamp.

1. Saw 2 pieces from 1" × 4" plate to 1-9/16" length.
2. Mill the sawed surfaces to 1.450".
3. Stack the 2 pieces to make the 2.900" dimension and clamp in a vise.

20   Chapter 1

4. Lay out, center drill and drill two 5/16" dia. holes 3" apart through both pieces.
5. Separate the two pieces and tap 3/8 × 16 thread 1" deep in the bottom plate.
6. Enlarge the two holes in the top plate to 13/32" dia.
7. Counterbore the top plate 9/16" dia. × 5/16 deep.
8. Bolt the two plates together with 0.005" shim.
9. Locate the center of the two plates. Center drill, drill to 1" dia. and bore to $2.000 ^{+000"}_{-001}$.
10. Drill and tap the bottom hole 3/8 × 16 thread.
11. Saw a 9/16" length from 1" × 1/2" bar.
12. Mill the 9/16" to 1/2" length.
13. Lay out and drill the 3/8" hole.
14. Drill and tap the 3/8" × 16 hole.
15. Mill the 1/16" slot.
16. Hand grind the left end to 1/4" radius.

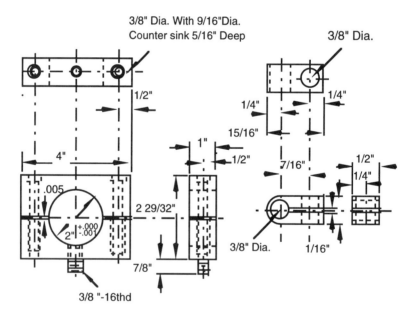

Figure 1-6. Clamp to hold the 3/8" diameter shaft of a dial indicator.

Materials and Manufacturing Considerations in Product Design

It took the skilled machinist, who was familiar with all of the tools in the shop, more than 12 hours to fabricate this clamp.

Rule 11. Evaluate the Preliminary Design by Considering Changes in the Configuration that Will Simplify the Fabrication

Then go through the design rules again. Repeat the process until the optimal design is achieved.

By applying this rule to the dial indicator clamp shown in Figure 1-6, the author realized that the original design was much too costly. Since the dial indicator weighed only about one pound, the stresses on the clamp would be negligible. Also, it is much simpler to use a V-shaped surface than a cylindrical surface to clamp onto a cylinder. Furthermore, forming a material, or a combination of forming and welding followed by machining, is in many cases much more economical than full machining.

The author then went down to the shop and fabricated the clamp shown in Figure 1-7. The operations were as follows:

1. Two pieces were cut off a 1/8" × 3/4" bar of steel; one 6-3/4" long and one 6" long.
2. Each piece was given one 90° and two 45° bends in a vise with a hammer.
3. 1/4" holes were drilled as noted on the drawing.
4. Three 3/16" × 1/2" long bolts with wing nuts were used to clamp the two pieces together.

The total time to fabricate this clamp was less than one hour.

22   Chapter 1

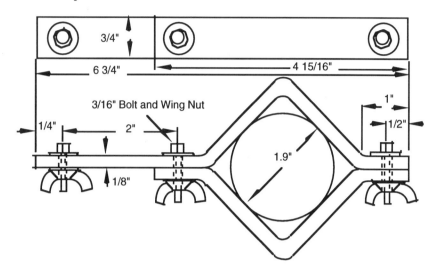

Figure 1-7. Redesigned dial indicator clamp.

The important lesson here is that when a drawing is sent to the shops, the drawing dictates the manufacturing processes that must be used to make the part — and what the final cost will be. The technicians on the shop floor do not redesign the part for ease of fabricability. Whether a part is made by machining, as shown in Figure 1-6, or by forming, as shown in Figure 1-7, is determined by the design engineer.

Materials and Manufacturing Considerations in Product Design

REFERENCES

1. Dixon, J.R. and C. Poli, *Engineering Design and Design for Manufacturing*, Field Stone, Conway, MA, 1995.
2. Datsko, J., *Material Properties and Manufacturing Processes*, John Wiley & Sons, New York, 1966.
3. Datsko, J., *Manufacturing Engineering Considerations In Product Design*, Proc. of 4th International Conference on Production Research, Tokyo, Japan, 1977.
4. Datsko, J. Materials Engineering Education for Design, Manufacturing and Materials Engineers, *Journal of Materials Engineering Education*, Vol. 10, 1988.
5. Helander, M. and M. Nagamachi, *Design for Manufacturability*, Taylor and Francis, London, 1992.
6. Lascoe, O.D. *Handbook of Fabrication Processes*, ASM International, Ohio, 1988.
7. Eary, D.F. and E.A. Reed, *Techniques of Pressworking Sheet Metal*, Prentice-Hall, New Jersey, 1978.
8. Budinski, K.G. *Engineering Materials: Properties and Selection*, Prentice-Hall, New Jersey, 1978.
9. Ruskin, A.M. *Materials Consideration in Design*, Prentice-Hall, New Jersey, 1967.

Chapter 2

THE STRUCTURE OF SOLIDS

INTRODUCTION

A study of the basic mechanical properties of materials must begin with an understanding of the structure of solid materials. In this context structures refer to the atomistic and crystalline patterns of which the solid material is composed. The definitions of the mechanical properties given in Chapter 7 are based on the crystalline structure of materials. For example, strength (and hardness) is defined as a material's ability to resist slip along its crystallographic planes. In order to increase the strength of a material, something must be done to it which will make slip more difficult to initiate. The following chapters explain the manner in which the various thermal and mechanical processes affect a material's structure, which in turn determines its mechanical properties. This chapter presents a brief review of atomic structure.

ATOMIC BONDING FORCES

The smallest particles that must be considered in the above context are atoms. The manner in which atoms are arranged in a solid material determines its crystal structure. The crystal structure and the type of interatomic bonding forces determine the strength and ductility of the material.

The simple model of an atom is a dense *nucleus*, consisting of *protons* and *neutrons*, surrounded by discrete numbers of planetary electrons orbiting in shells at specific distances from the nucleus. Each proton has an electrical positive charge of unity (1+). The number of protons in the nucleus determines the nuclear charge of the atom and is called the *atomic number*. Neutrons have no charge, but they do have mass. The *atomic weight* of an atom is the sum of the number of protons and neutrons. The electrons have negligible mass and a negative charge of unity (1–). The number of electrons in a given type of atom is also equal to the atomic number of that element. The maximum number of electrons in any shell or orbital is $2n^2$, where n is the quantum number of the shell. Thus, the maximum number of electrons that can be present in the first (innermost) shell is 2; in the second shell the maximum is 8. However, no more than 8 electrons are ever present in the outermost shell of an atom. The *valence* of an element is either the number of electrons in its outermost shell or the number of electrons necessary to fill the shell, whichever number is lower.

The interatomic bonding forces are determined by the valence, or outer shell, electrons. There are four types of atomic bonding forces that hold the atoms of a solid material in their relatively fixed positions. The three strongest (*ionic*, *covalent*, and *metallic*) types of bond are referred to as *primary*; the fourth type (*molecular*) is referred to as a *secondary* bond.

Ionic Bonds

From the above brief description of atomic structure it is evident that the uncombined atom is electrically neutral — the number of protons (+ charges) in the nucleus exactly equals the number of electrons (– charges). When atoms combine, only the valence electrons are involved. When a metal combines with a nonmetal, each metal atom "loses" its valence electrons and thus acquires a positive charge that is equal to the number of electrons so lost.

Likewise, each nonmetallic atom "gains" a number of electrons equal to its valency and acquires an equal negative charge. While in this state, the positively charged metallic atom and the negatively charged nonmetallic atom are called *ions*.

Like-charged particles repel each other and oppositely charged particles attract each other with an electrical force called the *coulomb force*. When a material is maintained in the solid state by the mutual attraction of positively and negatively charged ions, the interatomic bonding force is called ionic.

Ionic bonding is illustrated in Figure 2-1. The sodium atom (atomic number 11 with 11 electrons) and the chlorine atom (atomic number 17 with 17 electrons) are shown in (a). Since the number of electrons in both cases equals the number of protons (atomic number) in the nucleus, atoms are electrically neutral. When the sodium atom loses its valence electron, it takes on the stable electronic configuration of the inert neon atom, but, since it has one excess proton in the nucleus, it also takes on a positive charge. Likewise, the chlorine atom, upon gaining one electron in its outer shell, has the stable electronic configuration of the inert argon atom. Since the chlorine ion is deficient one proton, it has a negative charge. This condition is depicted in Figure 2-1(b), where the resulting charged ions are mutually attracted.

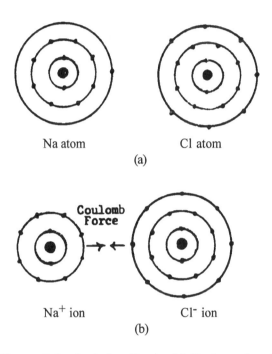

Figure 2-1. Ionic bonding. (a) Uncharged sodium and chlorine atoms exhibit no attractive force. (b) The + charged sodium ion and − charged chlorine ion are mutually attracted by coulomb forces.

As illustrated in Figure 2-1, ionic bonding by the two ions "sharing" an electron does not occur. If this were true, the sodium and chlorine ions would join up as pairs as is the case when atoms of a gas, such as hydrogen or oxygen, combine to form molecules. The ionic bond is nondirectional. That is, each positive ion attracts all neighboring negative ions and each negative ion attracts all neighboring positive ions. Thus, a sodium ion attracts as many

28   Chapter 2

chlorine ions around itself as it can to make the solid compound NaCl, whose cubic crystal structure is shown in Figure 2-2. The coulomb force attraction of oppositely charged ions is very large: therefore ionic bonded solids exhibit very high strength and relatively high melting temperatures. However, they exhibit very low ductility under normal conditions since interatomic bonds must be broken in order for the atoms to slide past each other. This is one of the most important distinctions between the ionic (and covalent) bonding and metallic bonding.

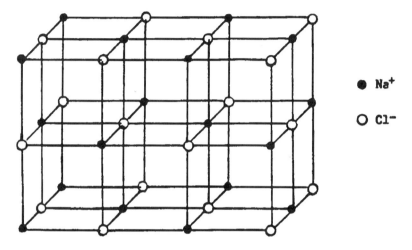

Figure 2-2. Cubic crystal structure of sodium chloride.

Covalent Bonds

Covalent bonds are those in which the atoms reach a stable configuration (filled outer shell) by *sharing* the valence electrons. Unlike the nondirectional ionic bonds, covalent bonds act between specific pairs of atoms and thus form molecules. Covalent bonds are most prevalent in gas molecules, as shown in Figure 2-3.

Gas molecules of one type of atom, such as hydrogen, oxygen, nitrogen, and fluorine, are electrically symmetrical and neutral, as can be seen in both the single- and the double-bonded molecules of H_2, Cl_2 and O_2 (Figure 2-3). These primary bonds are very strong, as evidenced by the difficulty in dissociating these molecules.

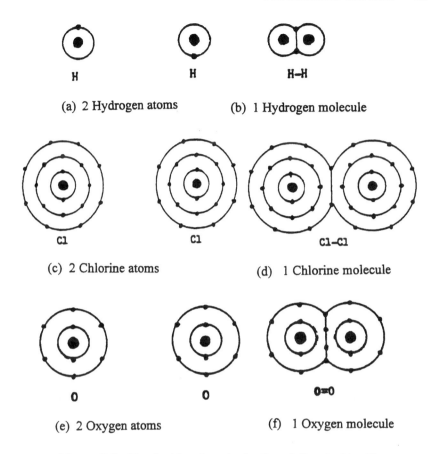

Figure 2-3. Covalent bonds: single (b and d); double (f).

Covalent bonding also results in the formation of very large molecules, which are present as solids rather than as liquids and gases. Diamond, silicon, and silicon carbide are examples of such covalently bonded solids. They are characterized by high strength and melting temperature and by low ductility. The atoms in a diamond structure are arranged on two interpenetrating face-centered cubic lattices, as shown in Figure 2-4. The entire crystal is composed of only one molecule. In order to fracture the crystal the strong covalent interatomic bonds must be broken.

Figure 2-4. Diamond structure. The entire crystal is one large molecule. Each carbon atom has 4 bonds.

Metallic Bonds

Of the three primary bonding forces, the metallic bond is by far the most important to an understanding of the mechanical properties of the materials with which the practicing engineer is concerned. Unfortunately, it is also the most difficult to describe in elementary terms since the correct model of atomic bonding can be presented only in terms of quantum mechanics and wave mechanics, which are beyond the scope of this book. However, without going into the mathematics of atomic physics, the following simple model of metallic bonding is presented to explain the properties of metals, which constitute the majority of the elements and are the focus of materials design and manufacturing.

The *metallic bond* is a special type of covalent bond wherein the positively charged nuclei of the metal atoms are attracted by electrostatic force to the valence electrons that surround them. Unlike the common covalent bond which is directional, i.e., between a pair of atoms, the metallic bond is nondirectional and each nucleus attracts as many valence electrons as possible. This leads to a dense packing of the atoms, and thus the most common crystal structures of the metal are close-packed: face- and body-centered cubic and hexagonal close-packed.

Metal atoms have their own unique type of bonding force because of the looseness with which their valence electrons are held in the outer shell. This is evident from the fact that the ionization potential of the metal atoms is 1/2 to 2/3 that of the nonmetal atoms. The mean radius of the valence electrons is closer to a nucleus in the solid metal than they are in a free atom and thus their potential energy is lower in the solid.

Since the valence electrons are not localized between a pair of positive ions, they are free to move through the solid. Thus the structure of the solid metal is a close-packed arrangement of positive ion "cores" (the nucleus plus the nonvalence electrons) that is permeated by an electron "gas" or "cloud."

This ability of the valence electrons to move freely through a solid explains the high thermal and electrical conductivities of metals. Also, the fact that the valence electrons are nondirectional (not shared by only two atoms) explains the low strength and high ductility of elemental metals since the positive ions can move relative to one another without breaking any primary bonds. This mechanism is referred to as *slip* and is discussed in more detail in the section on crystal structures. Figure 2-5 shows a schematic representation of the metallic bond. In this structure the positive ions repel each other, as do the electrons, but the ions and the electrons mutually attract each other. The resultant of these repulsive and attractive forces is an equilibrium distance between the positive ions that determines the lattice parameter (side length of the unit cell). Tensile forces (or a rise in temperature) must be exerted to extend the lattice and compressive forces (or a reduction in temperature) to contract it.

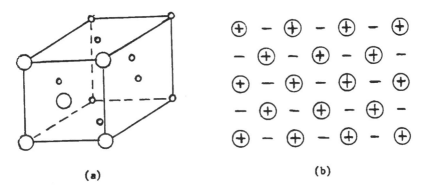

Figure 2-5. Metallic bonding. A unit cell of a face-centered cubic metal is shown in (a) with large circles designating the front face. The front face only of 6 adjacent unit cells is shown in (b) with the valence electrons (–) dispersed. The electrons are not actually in the plane of the ions.

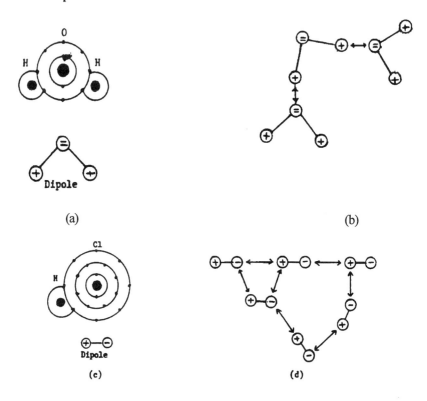

Figure 2-6. Secondary molecular bonds. (a) All covalent bonds in a water molecule. (b) Three water molecules attracted by van der Waals attractive forces (arrows). (c) An HCl molecule. (d) HCl molecules attracted by van der Waals forces.

Molecular Bonds or van der Waals Bonds

In addition to the three strong primary bonds discussed above, there are several much weaker (and therefore called secondary) bonds which provide the interatomic attractive forces that hold some types of atoms together in a solid material. These forces are referred to as either *secondary bonds*, *molecular bonds*, or *van der Waals bonds*. These bonds are due to residual electrostatic fields between electrically neutral molecules whose charge distribution is not uniform. van der Waals forces were first used to explain the "bonding" of atoms in electrically neutral gas molecules.

Covalently bonded atoms frequently form molecules that behave as electrical or magnetic *dipoles*. Although the molecule itself is electrically neutral, there is an electrical imbalance within the molecule: that is, the center of the positive charge and the center of the negative charge do not coincide. Figure 2-6 shows how a nonsymmetrical molecule produces a dipole, and how this dipole creates molecular bonding.

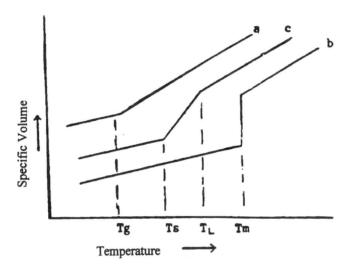

Figure 2-7. Specific volume vs. temperature. (a) Glass with a transition temperature T_g. (b) A crystal that melts at a fixed temperature T_m, such as a pure element compound. (c) A crystal that melts over a range of temperature, such as a solid solution alloy, with T_l the liquidus temperature and T_s the solidus temperature.

ATOMIC STRUCTURE

Whereas the electrical properties of a material depend upon the internal structure of the atoms, the mechanical properties of a material depend upon the type of structure that groups of atoms form. In this context, *atomic structure* refers to the arrangement of atoms and not to the internal structure of individual atoms. All solid materials can be classified into one of three groups (in order of increasing effect on mechanical properties) based on atomic structure: amorphous, molecular or crystalline. Knowledge of the atomic structure of solids makes it possible to understand why a given material has its unique properties and what condition it should be in to achieve optimal mechanical properties.

Amorphous Solids

Amorphous materials lack repetitive arrangement of the constituent atoms. In an atomic sense they have no "structure." Gases and liquids are amorphous materials: the most important amorphous solids are the glasses, and they are frequently considered simply supercooled liquids.

Glass behaves as a typical liquid at high temperatures. The atoms are very mobile and do not vibrate in a fixed location in space. A given mass of hot glass, like any other liquid, takes the shape of the container in which it is placed.

As a hot glass cools, the atoms vibrate at lower amplitudes and come closer together, resulting in an overall thermal contraction or decrease in specific volume. The decrease in specific volume of a liquid as temperature decreases is approximately linear and occurs with all liquids including liquid metals. This is illustrated in Figure 2-7.

When the hot liquid glass is cooled to the *glass transition temperature*, T_g, there is an abrupt change in the slope of the cooling curve. Below the T_g temperature, glass behaves as a typical solid.

Molecular Solids

A *molecule* is a group of atoms that are held together by strong ionic or covalent bonds. A *molecular solid* is a structure made up of molecules that are attracted to each other by the weak van der Waals forces. The two most common types of molecular solids are the silicates and the polymers. The silicates have ionic intramolecular bonds and the polymers have covalent ones. The polymers, which are more important in terms of mechanical properties in manufacturing, will be discussed in more detail.

The Structure of Solids 35

Polymers are organic compounds of carbon, hydrogen, and oxygen to which other elements such as chlorine or fluorine may be added. They cover a wide range of structural arrangements with resulting variations in properties. Large molecules are constructed from a repeating pattern of small structural units. The hydrocarbons have repeating structural units of carbon and hydrogen atoms. The smallest or simplest unit is four atoms of hydrogen attached to one of carbon; its chemical formula is CH_4 and it is called *methane*. Figure 2-8 shows both the three-dimensional structure of the methane molecule and its two-dimensional representation. The four hydrogen atoms are at the four corners of a tetrahedron. To see the tetrahedron in Figure 2-8(a) it is necessary to draw the six lines that connect the four atoms. These lines are omitted in the figure for clarity.

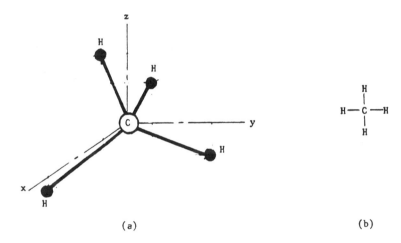

Figure 2-8. Methane molecule. (a) Three-dimensional representation with the carbon atom at the center and the four hydrogen atoms at the four corners of a tetrahedron. (b) Two-dimensional representation.

Figure 2-9 illustrates how larger molecules are created by extending CH_2 links onto the basic CH_4 (methane) building unit. These chainlike series of hydrocarbon molecules are called *paraffins* and they fit the formula C_nH_{2n+2}. These molecules are referred to as *saturated* because they contain only single bonds. (*Unsaturated* molecules contain double or triple bonds). Paraffin molecules have very strong covalent intramolecular bonds but only weak intermolecular van der Waals bonds. The strength of the van der Waals forces, and consequently the melting temperature, increases as the size of the molecule increases. Molecules containing 4 or fewer carbon atoms have such low secondary bonding forces that they normally exist as gases. Molecules containing 5 to 16 carbon atoms have sufficiently high secondary bonds that they normally exist as liquids of increasing viscosity. The van der Waals forces are large enough to attract molecules having more than 16 carbon atoms into what is normally a solid structure. The household wax called paraffin contains approximately 30 carbon atoms and melts at approximately 100°F. Polyethylene contains several hundred carbon atoms and has a melting temperature that is close to the 300°F maximum value for the paraffin series.

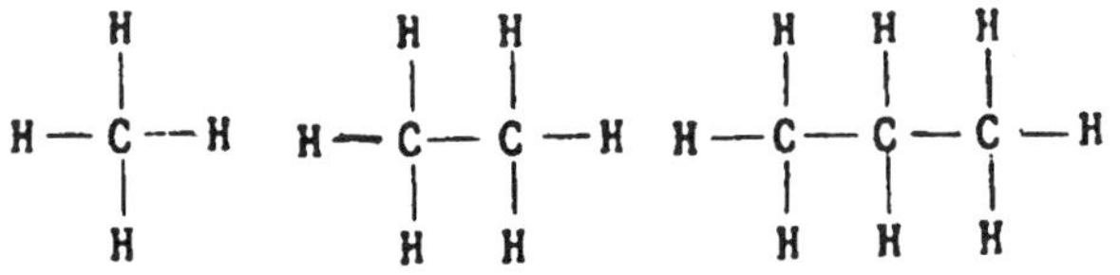

CH_4
(a) Methane

$+ CH_2$
(b) Ethane

$+ CH_2$
(c) Propane

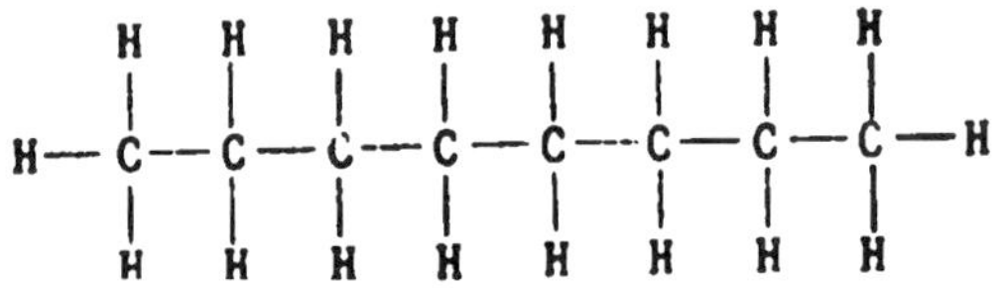

$+$ five CH_2
(d) Octane

Figure 2-9. Increasingly large, saturated (single-bonded) molecules formed from methane.

Some molecules of a given composition may exist in two or more structural forms, each having somewhat different properties owing to their having different molecular dipoles. Molecules of this type are called *isomers*. Figure 2-10 illustrates the two structures of butane, C_4H_{10}, which is the smallest paraffin molecule that can exist in more than one structural form. All larger paraffin molecules are isomers.

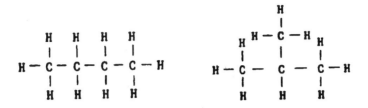

Figure 2-10. Butane isomers.

Polymerization The most important polymeric solid materials are large molecules (macromolecules) that are built up from small, unsaturated molecules called *monomers* by the polymerization processes of addition or condensation. In order for a molecule to be classified as a monomer, it must be *polyfunctional*. That is, it must contain two or more reaction sites such as double bonds, hydroxyl radicals, amino acids, or polyalcohols. With two reaction sites it is bifunctional; with three it is trifunctional. In the *addition* type of polymerization, the double bonds of a monomer are broken into single bonds and the individual molecules become connected to each other. Thus, a large molecule is formed without any by-products. In the *condensation* type of polymerization, the monomers combine to form a large molecule plus a by-product, which is a simple molecule such as H_2O or HCl.

Figure 2-11 shows some of the more common monomers or unsaturated molecules that are used in the building of macromolecules. The simplest monomer is ethylene, C_2H_4, and is shown in Figure 2-11(a). It is the base in the group of hydrocarbons called *olefins*. The olefins have the chemical formula C_nH_{2n}. The benzene molecule shown as (d) in Figure 2-11 is another important building unit. Because of the shape of the molecule, it is described as a ring molecule or compound. The benzene group is also called the *aromatic* hydrocarbons.

38   Chapter 2

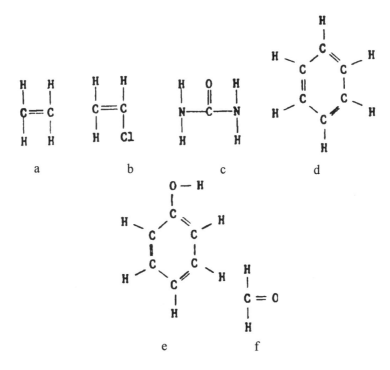

Figure 2-11. Monomers. Small, unsaturated (double-bonded) molecules that are building units for larger polymer molecules: (a) ethylene, (b) vinyl chloride, (c) urea, (d) benzene, (e) phenol, (f) formaldehyde.

Figure 2-12 illustrates the addition polymerization of the ethylene monomer. The double bonds of ethylene are broken in the presence of a catalyst such as boron trifluoride. The vinyl chloride monomer as shown in Figure 2-11(b) is similar to ethylene except that one of the hydrogen atoms is replaced with a chlorine atom. The polymerization of this monomer results in polyvinyl chloride. These macromolecules resemble smooth strings or chains, as can be seen from their structural arrangement.

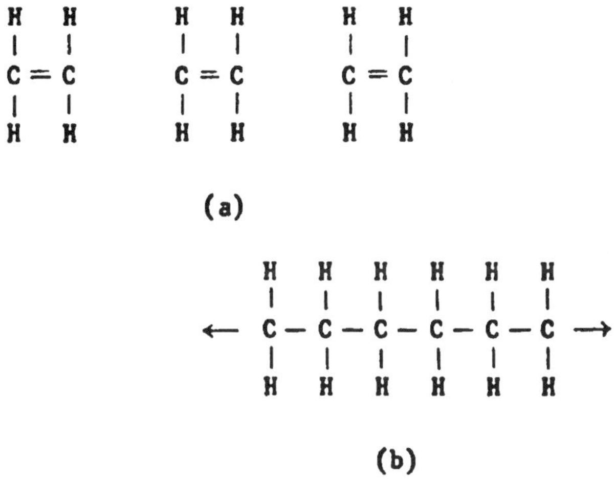

```
H   H      H   H      H   H
|   |      |   |      |   |
C = C      C = C      C = C
|   |      |   |      |   |
H   H      H   H      H   H
```

(a)

```
        H   H   H   H   H   H
        |   |   |   |   |   |
   ← C - C - C - C - C - C →
        |   |   |   |   |   |
        H   H   H   H   H   H
```

(b)

Figure 2-12. Addition polymerization. (a) Three individual monomers of ethylene. (b) A portion of a polyethylene molecule formed when each double bond of the monomers is broken by a catalyst to form two single bonds and join the individual molecules together.

Some macromolecules resemble rough chains; that is, chains with many short side arms branching from it. Polystyrene, which is a very important industrial polymer, is of this type. The styrene monomer is made from the benzene ring (C_6H_6) with one of the hydrogen atoms replaced with a $CH = CH_2$ molecule, as shown in Figure 2-13(a). Polymerization then occurs by breaking the double bond in the $CH = CH_2$ group with the help of a peroxide catalyst and joining the two of them as shown in (b).

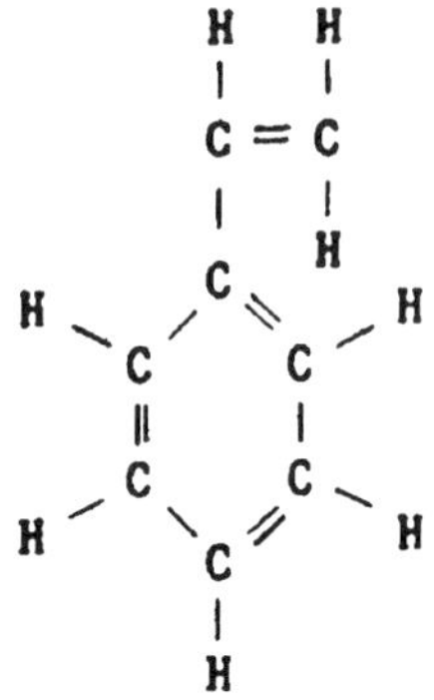

Figure 2-13. (a) Styrene and (b) polystyrene structures. The polymerization takes place in the presence of a peroxide catalyst.

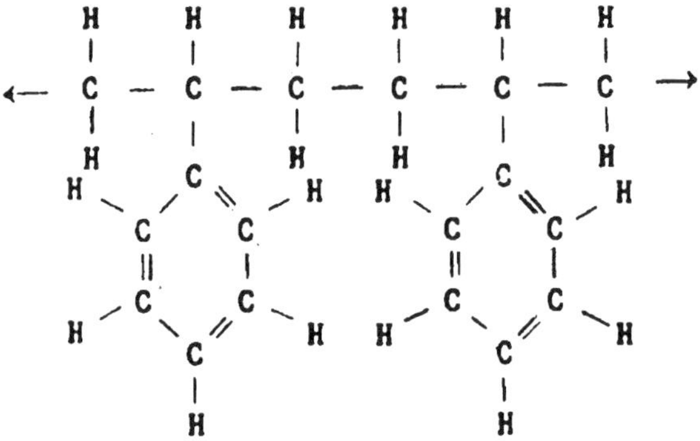

All of the polymers described above are *thermoplastic*: they melt or soften when they are heated. This is because the individual macromolecules are stable but the linkages to other macromolecules are loose (since they are attracted to each other only by weak van der Waals forces). Some polymers are *thermosetting*: they do not soften when they are heated but retain their "set" or shape until charred. This is because the individual macromolecules unite with each other and form many cross linkages. Bakelite (phenal formaldehyde) is such a polymer. Figure 2-14 shows how each formaldehyde monomer joins two phenol monomers together, under suitable heat and pressure, to form a macromolecule. This is a condensation type of polymerization because one water molecule is formed from the oxygen atom of each formaldehyde molecule and the hydrogen atoms from each of the two phenol molecules.

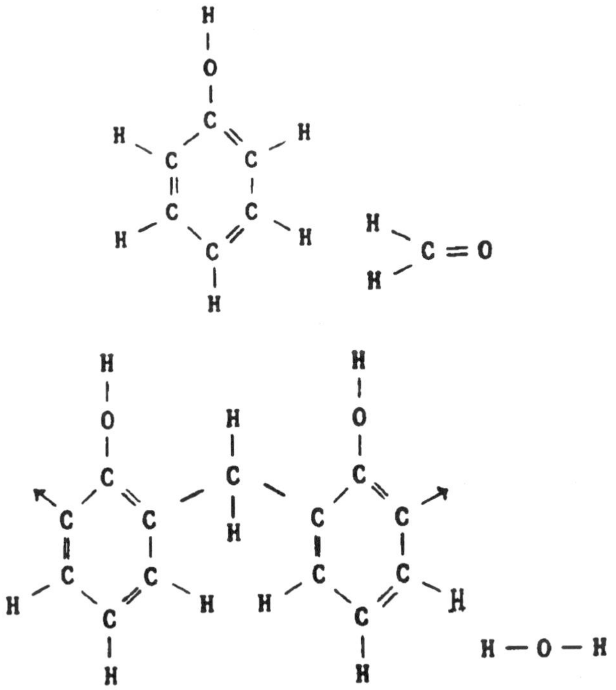

2 phenol + 1 formaldehyde molecules = phenol formaldehyde + H_2O (under heat and pressure)

Figure 2-14. Condensation polymerization of phenol and formaldehyde into Bakelite.

Mechanical properties of molecular structures The mechanical properties of polymers are determined by the type of forces acting among the molecules. The polymers are amorphous, with random chain orientations while in the liquid state. This structure can be retained when the polymer is cooled rapidly to the solid state. In this condition the polymer is quite isotropic. However, with slow cooling or plastic deformation, such as stretching or extruding, the molecules can become aligned. That is, the chains of all the molecules tend to become parallel along the long axis. A material in this condition is said to be *oriented* or *crystalline*, the degree of orientation being a measure of the crystallinity. When the molecular chains of a polymer have this type of directionality, the mechanical properties are also directional and the polymer is anisotropic. An aligned polymeric material is stronger along the axis of the chains and much less strong in the perpendicular directions. This is because the individual aligned macromolecules are attracted only by weak van der Waals forces, whereas the atoms along the axis of the chains are held together by strong covalent bonds. The intermolecular strength of the linear polymers can be increased by the addition of polar (dipole) groups along the length of the chain. The most frequently used polar groups are chlorine, fluorine, hydroxyl, and carboxyl.

Thermoset (cross-linked) polymers have all of the macromolecules connected in three directions by strong covalent bonds. Consequently, these polymers are stronger than the thermoplastic ones, and they are also more isotropic.

Crystalline Solids

Crystalline solids are used extensively in manufacturing because of their mechanical properties and load-carrying capacity. Of all the crystalline solids, the metals are the most important. But before one begins a study of the properties of metals, it is necessary to review some of the basic concepts of crystal structure. The structural arrangement of crystals can be studied by x-ray diffraction analysis, which has resulted in the following models of crystal structures.

A *crystal* (or crystalline solid) is an orderly array of atoms having a repeating linear pattern in three dimensions. The atoms are represented as spheres of radius r. A *space lattice* is the three-dimensional network of straight lines that connect the centers of the atoms along three axes. The intersections of the lines are lattice points and they designate the locations of the atoms. Although the atoms vibrate about their centers, they occupy the fixed positions of the lattice points. Figure 2-15 shows a space lattice with the circles representing the centers of the atoms. A space lattice model has two important characteristics: (1) the space lattice network divides space into equally sized prisms whose faces contact each other in such a way that no void spaces are present; (2) every lattice point of a space lattice has identical surroundings.

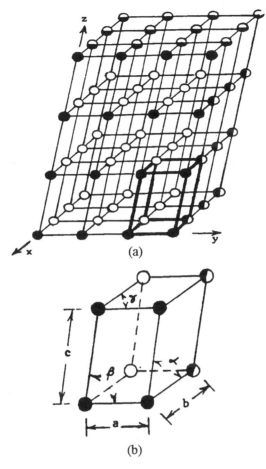

Figure 2-15. A space lattice. (a) A unit cell is marked by the heavy lines. Black circles are on the front face; horizontal shading on the top face; vertical shading on the right-side face; hidden circles are white. (b) An isolated unit cell showing dimensions a, b, c and angles α, β, γ.

The individual prisms that make up a space lattice are called *unit cells*. A unit cell is the smallest group of atoms which, when repeated in all three directions, make up the space lattice, as illustrated by the dark-lined parallelopiped of Figure 2-15. Each unit cell in a space lattice has the same size, shape, and orientation. The size and shape are specified by the three side lengths (a, b, c) and the three angles (α, β, γ), as shown in Figure 2-15.

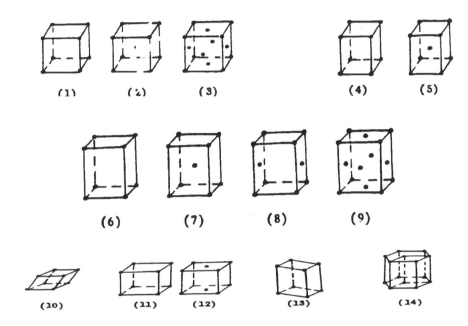

Figure 2-16. The 14 space lattices: (1) simple cubic, (2) body-centered cubic, (3) face-centered cubic, (4) simple tetragonal, (5) body-centered tetragonal, (6) simple orthorhombic, (7) body-centered orthorhombic, (8) end-centered orthorhombic, (9) face-centered orthorhombic, (10) rhombohedral, (11) simple monoclinic, (12) end-centered monoclinic, (13) triclinic, and (14) hexagonal.

Only 14 different space lattices and 7 different systems of axes are possible. These are illustrated in Figure 2-16 and listed in Table 2-1. Most of the metals belong to three of the space lattice types: face-centered cubic, body-centered cubic, and hexagonal close-packed. They are listed in Table 2-2, along with four metals that have rhombohedral and two that have orthorhombic structures. Although the rhombohedral metals and one of the orthorhombics have low melting temperatures, there is no obvious relationship between type of crystal structure and melting temperature. For example, gallium, which melts at 85°F, and iron carbide (cementite), which has no melting temperature but sublimes at a very high temperature, both have orthorhombic structures.

Table 2-1. Crystal Systems and Space Lattices

Crystal System	Space Lattice	Number of Axes	Side Length	Interaxial Angle
Cubic	Simple Body-centered Face-centered	3	$a = b = c$	$\alpha = \beta = \gamma = 90°$
Tetragonal	Simple Body-centered	3	$a = b \neq c$	$\alpha = \beta = \gamma = 90°$
Orthorhombic	Simple Body-centered Base-centered Face-centered	3	$a \neq b \neq c$	$\alpha = \beta = \gamma = 90°$
Rhombohedral	Simple	3	$a = b = c$	$\alpha = \beta = \gamma \neq 90°$
Monoclinic	Simple Base-centered	3	$a \neq b \neq c$	$\alpha = \beta = 90° \neq \gamma$
Triclinic	Simple	3	$a \neq b \neq c$	$\alpha \neq \beta \neq \gamma \neq 90°$
Hexagonal	Simple	4	$a_1 = a_2 = a_3 \neq c$ or $a_1 = b \neq c$	$\alpha = \beta = 90°$ $\gamma = 120°$

The crystalline structure is not restricted to metallic bonding; ionic and covalent bonding are also common. The metallic-bonded crystals are very ductile because the valence electrons are not associated with specific pairs of ions.

Face-centered cubic Most of the common metals (see Table 2-2) have face-centered cubic (FCC) structures. Figure 2-17(a) shows the arrangement of the atoms, represented by spheres, in the FCC structure as well as that fraction or portion of each atom associated with an individual unit cell. Each atom in the FCC structure has 12 contacting atoms. The number of contacting atoms (or nearest neighbors) is called the *coordination number*. The relationship between the lattice parameter *a* and the atomic radius *r* can be determined from Figure 2-17(b) as

$$a = 2r\sqrt{2} \qquad (2\text{-}1)$$

Table 2-2. Lattice Structure of Metal Crystals

Face-centered cubic		Body-centered cubic		Hexagonal close-packed		Rhombo-hedral	Ortho-rhombic
Ag	Ni	Cb	Li	Be	Se	As	Ga
Al	Pb	α Cr	Mo	Cd	Te	Bi	U
Au	Pd	Cs	Na	α Co	Ti	Hg	
Ce	Pt	α Fe	Ta	β Cr	Tl	Sb	
β Co	Rh	δ Fe	V	Hf	Y		
Cu	Sc	K	W	Mg	Zn		
γ Fe	Th			Os	Zr		
Ir	βTi			Ru			

As an example, the aluminum atom has a diameter of 2.862 $\overset{o}{A}$, where $\overset{o}{A}$ is the symbol for angstrom units. An angstrom unit equals 10^{-8} cm. Thus the side length of the unit cell is $2.862\sqrt{2} = 4.0414 \overset{o}{A}$, or 4.0414×10^{-8} cm.

The density of a crystalline solid can also be calculated from Figure 2-17 and the atomic weight of the element. For example, aluminum has an atomic weight of 26.97 g/g atomic weight. Since all materials have 6.02×10^{23} atoms /g atomic weight, aluminum has a weight of 4.48×10^{-23} g/atom. The volume of the unit cell containing 4 atoms is 4.04143×10^{-24} cm^{-3}. Thus the density (weight/unit volume) of aluminum is 4 atoms $\times 4.48 \times 10^{-23}$ g/atom divided by 4.04143×10^{-24} cm^{-3} or .71 g/cm^{-3}. This is in agreement with the experimentally determined value.

The FCC structure is referred to as a dense or closely packed structure. A quantitative measure of how efficiently the atoms are packed in a structure is the *atomic packing factor* (APF) which is the ratio of the volume of the atoms in a cell to the total volume of the unit cell. The volume of a sphere is $1.333 \pi r^3$ The volume of an FCC unit cell is a^3, or, from Eq. (2-1), $22.64 r^3$.

Thus the APF is $4 \times 1.333\pi r^3 / 22.64 r^3$, or 0.74. This means that 26% of the FCC unit cell is "void" space. It is obvious that the atomic packing factor depends only upon the crystal structure and is independent of the atomic radius. Thus all FCC structures have 26% void space.

Although a simple cubic is one of 14 possible structures, none of the materials has this lattice shape. The reason for this is apparent from the previous discussion of bonding forces in which it was pointed out that the ions tend to form close-packed arrangements. But the APF for the simple cube is $1.333\pi r^3 / (2r)^3$, or 0.52. Thus the simple cubic structure has 48% "void" space, which is nearly twice as much as the FCC structures has. Since the simple cubic structure is such an inefficient arrangement of atoms, it does not occur frequently. Some crystal structures, such as sodium chloride (see Figure 2-2), appear to be simple cubic, whereas they are really two interpenetrated FCC lattices. The APF for the NaCl lattice is 0.67.

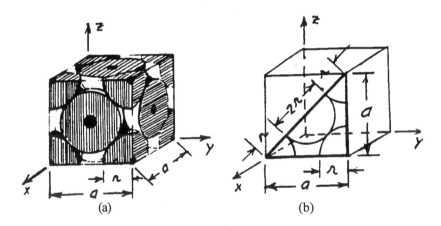

Figure 2-17. Unit cell of face-centered cubic structure. (a) The unit cell has 8 corners with 1/8 atom at each, plus 6 faces with 1/2 atom, for a total of 4 atoms per unit cell. (b) One half of the front face showing the relationship between the lattice parameter a and the atomic radius r.

Body-centered cubic Many of the stronger metals (Cr, Fe, Mo, W) have body-centered cubic (BCC) lattice structures while the softer more ductile metals (Ag, Al, Au, Cu, Ni) have the FCC structure (see Table 2-2). Figure 2-18(a) shows

48   Chapter 2

the arrangement of the atoms in the BCC structure. There are two atoms per unit cell: one in the center (body center) and 1/8 in each of the 8 corners. As can be seen in Figure 2-18(a), each atom, being contacted by 8 other atoms, has a coordination number of 8.

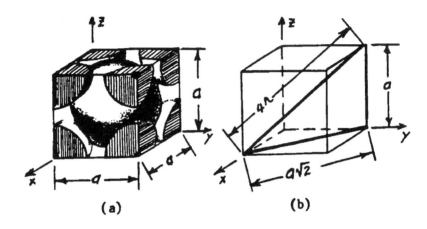

Figure 2-18. Unit cell of body-centered cubic structure. (a) The unit cell has 1/8 atom at each of 8 corners and 1 atom at the geometric center of the cell, for a total of 2 atoms. (b) The relationship of the lattice parameter a and atomic radius r.

The relationship between the lattice parameter a and the atomic radius r can be determined from the triangle sketched in Figure 2-18(b).

$$a = 4r\sqrt{3} \qquad (2\text{-}2)$$

Thus α iron (the room temperature structure) has an atomic radius of 1.2389 Å, so that the side length of its unit cell is 2.861 Å.

The atomic packing factor for the body-centered cubic unit cell is 0.68, which is a little lower than the FCC. It is interesting to note in Table 2-2 that slightly fewer metals have the BCC structure compared with the FCC. The hexagonal close-packed (HCP) structure, which will be described next, also has an APF of 0.74 and a coordination number of 12, both being identical to the FCC structure. Table 2-2 also shows that the number of metals having the HCP structure is approximately the same as the number with the FCC structure.

Hexagonal close-packed The *simple hexagonal* structure, as shown in Figure 2-16(14), consists of two hexagons, one directly above the other, with atoms at the six corners and one in the center. It has a low APF and consequently no metals have this structure. However, if three atoms are placed in the center of the "void" spaces between the two hexagons, a more dense structure results, which is called the hexagonal close-packed.

Figure 2-19(a) shows the hexagonal unit cell that is one of the 14 basic lattice structures and Figure 2-19(b) the hexagonal close-packed modification of it. The hexagonal unit cell is made up of three identical rhombic prisms having base angles of 120° and 60°. The three additional atoms that allow for HCP structure lie at the "void" centers of the three prisms. This can be seen more clearly in the isolated prism of Figure 2-19(c).

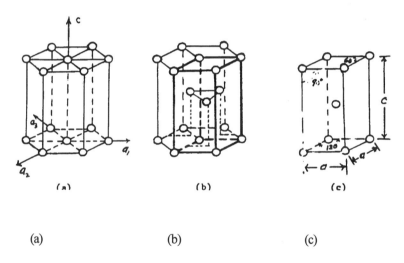

(a)   (b)   (c)

Figure 2-19. The hexagonal unit cell and the HCP modification. (a) The simple hexagonal unit cell. (b) The HCP structure. It is not one of the basic 14 lattices. (c) A cell of the HCP showing its location.

Many metals have this structure, with its high atomic packing factor (0.74). Because most of the atoms lie on the basal planes of the hexagonal system, plastic deformation can occur more readily along these planes than across them. Therefore, the HCP metals are very anisotropic.

50   Chapter 2

Crystal directions and planes   In order to conduct and communicate research on the properties of crystals, it is necessary to have a system of notation for both the directions in a crystal and the orientation of the crystallographic planes. Directions in a space lattice are represented by vectors.

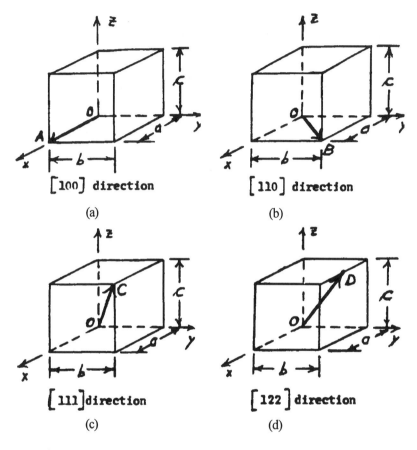

Figure 2-20. Direction indices. The unit cell has side lengths of a, b, and c in the x, y, and z lattice directions. The side lengths need not be equal and the angles need not be 90°.

The commonly accepted standard notation for directions in a crystal is based on multiples of the side lengths of the unit cell and is called the *direction indices*. These indices are the vector components of the direction ray resolved

The Structure of Solids 51

onto the coordinates axes and reduced to the smallest set of integers that are multiples of side lengths. This notation can best be demonstrated by several examples.

Figure 2-20(a) shows the direction ray, OA which lies on the x axis. To get to A from O, it is necessary to go one a distance along the x axis, zero b distance along the y axis, and zero c distance along the z axis. By dropping the unit cell lengths, the position of A with respect to O is 1 unit, 0 unit, and 0 unit in the x, y, and z direction, respectively. The direction indices are always given in the above order and within brackets to designate them as direction indices rather than crystallographic planes. Thus the ray OA is designated by the direction indices [100].

To get to point B from O in Figure 2-20(b), it is necessary to go one a distance along the x axis, one b distance along the y axis, and zero c distance along the z axis. Therefore, the direction ray OB, which is a face-diagonal, is defined by the direction indices [110]. Likewise, the body diagonal direction OC in Figure 2-20(c) has the direction indices [111].

If the direction ray is a fraction of a unit cell side length, then the direction indices must be reduced so that they can be expressed as a ratio of smallest integers. Thus the direction ray OD in Figure 2-20(d) has the position $1/2a$, $1b$, $1c$. By reducing each distance to the same denominator and dropping the side lengths, the position of D becomes 1/2, 2/2, 2/2. The smallest integers having the same proportions are 1, 2, 2. and so the direction indices of ray OD are [122]. Likewise, if a point E is located at $1/2a$, $1/3b$, and $2/3c$, its position can be restated as 3/6, 2/6, 4/6, and so its direction is designated as [324].

If a direction ray is more than a unit cell side length, the direction indices are again reduced to the smallest integers. Thus a point located at $2a$, $2b$, and $2c$ is on the line that goes through $1a$, $1b$, and $1c$, so it has the direction indices [111].

Negative directions (directions opposite to those shown in Figure 2-20) have a negative sign above the digit. Thus if point A in Figure 2-20(a) went in the opposite direction, its direction indices would be [$\bar{1}$00].

Several directions in a crystal are equivalent because of crystal symmetry. For example, the edges of a unit cell are all equivalent. A full set of equivalent directions is indicated by replacing the brackets with angle brackets, < >. Therefore the direction indices for all the edges of a unit cell are specified by <100>.

The *Miller indices* are used to designate specific crystallographic planes within a space lattice. They specify the orientation of the planes with respect to the axes of the unit cell. They do not fix the position in terms of distance from

52   Chapter 2

the origin: thus parallel planes have the same designation. The Miller indices are determined from the three intercepts that the plane makes with the three axes of the crystal. Actually, it is the reciprocal of the distance between the intercepts with the axis and the origin measured in terms of multiples of fractions of the unit cell lengths a, b, and c that is used in the determination. The final step in calculating the Miller indices is to reduce the three reciprocals to the lowest integers having the same ratio. The indices are expressed in parentheses. As is true with the direction indices, the sequence of the integers relates to the distances along the x, y and z axis, respectively. The following examples make this procedure clear.

Figure 2-21(a) identifies the front face of the crystal with the Miller indices (100). This notation is arrived at as follows. The front face intersects the x axis at one a distance and it does not intersect the y and z axes (or it intercepts at zero b and zero c units). If the side lengths are dropped, the intersects are 1, 0, 0. The reciprocals of these are also 1, 0, 0. Since these are already the smallest integers, the Miller indices are specified by enclosing them in parentheses, thusly (100). (The commas are omitted in the notation.)

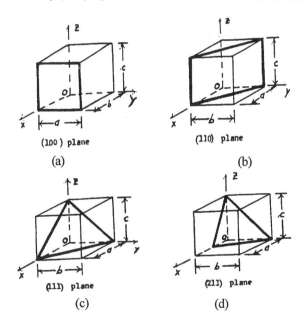

Figure 2-21. Miller indices for some crystallographic planes.

Figure 2-21(b) shows the (110) plane that is parallel to the z axis and is a face diagonal on the top and bottom faces of the unit cell. This plane intersects the x axis at 1a distance, the y axis at 1b distance, and the z axis at 0c distance. The intersects are 1, 1, 0; and so are the reciprocals. Since these are the smallest integers with which this ratio can be expressed, the Miller indices are specified as (110).

Figure 2-21(d) shows the crystallographic plane that intersects the x axis at 1/2a the y axis at 1b, and the z axis at 1c. The reciprocals are therefore 2, 1, 1, and so this plane is identified as the (211) plane. As is true with the direction indices, planes that intersect on the opposite side of the origin 0 (in the negative quadrant) have a negative sign placed above the integer.

When one is discussing the Miller indices in general, that is, when one is not referring to a specific plane, the integers are replaced with the letters h, k, l (in that order). Thus, in the notation (hkl), h represents the reciprocal of the x axis intersection.

Parentheses are used, as in the above examples, to specify a single plane or a family of parallel planes. Thus (100) represents all of the planes that are parallel to the yz axes and intersect the lattice structure at 1a, 2a, 3a, etc., distances. Wavy brackets, or braces, { }, are used to designate all planes in a crystal that are equivalent. For example, the six face planes of a unit cell such as in Figure 2-21(a) are: (100), (010), (001), ($\bar{1}$00), (0$\bar{1}$0), (00$\bar{1}$). The notation {100} includes all of these six planes.

Crystal Imperfections

The previous discussions of crystal structure have assumed a perfect lattice, that is, with an atom occupying each and every lattice point and the distance between equivalent lattice points being exactly the same. In the early 1900s it was found that real crystals did not have the properties theorized for perfect crystals. Notable was the discrepancy in mechanical strength, much lower than perfect crystals should have. As early as 1928 Prandtl suggested that slip (plastic deformation) in a crystal and the strength of a crystal are related to the presence of linear imperfections within the crystals. This type of imperfection is now called a dislocation. Today the terms *imperfection* and *defect* refer to a deviation from a perfectly ordered lattice structure.

Lattice imperfections are classified into three types: *point defects*, where the imperfection is localized about a single lattice point and involves only a few atoms; *line defects*, where the imperfection lies along a line of finite length and

54   Chapter 2

involves a row (line) or many atoms; and *plannar defects* or *boundaries*, where the imperfections involve entire planes of atoms such as the interface between adjacent crystals.

Point defects   Point defects are caused by (1) the absence of an atom from a lattice point, (2) the presence an extra atom (usually a small foreign one) in the "void" spaces of the lattice, (3) the presence of a foreign atom in one of the lattice sites, or (4) atoms that are displaced from their normal position in the array. Figure 2-22 illustrates these defects.

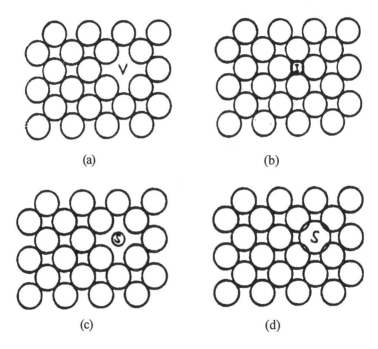

Figure 2-22. Some common point defects: (a) vacancy; (b) interstitial atom, I; (c) substitution of a smaller atom, S; (d) substitution of a larger atom, S.

The first type of point defect, the absence of an atom from a lattice point, is called a *vacancy*. Figure 2-22(a) shows a vacancy on the (100) planes of an FCC lattice. As depicted, the defect has been exaggerated in order to make it more apparent. In reality the neighboring atoms nearby the vacant site would be displaced from their normal position in the array to centers closer to the vacancy. Thus, the lattice lines joining the centers of the atoms are not straight in the vicinity of the vacancy. In three dimensions this means the crystallographic planes are warped inward near a vacancy. Individual vacancies can cluster together to form larger voids. Vacancies have no effect on the metallurgical mechanical properties discussed in the later chapters. However, they do affect properties such as conductivity and diffusivity.

The second type of point defect, the presence of an extra atom at the interstices of the lattice, is known as an *interstitial defect*. This type of defect in the lattice structure can be corrected by interstitial alloying, in which the solute atom fits in the interstices of the solvent lattice. This strengthening mechanism accounts for the high strength of fully hardened (heat-treated) steel. Commercially pure iron (ferrite with a BCC structure) has a yield strength of 10 to 20 ksi. However, with 0.8% carbon dissolved interstitially in the iron, and stress relieved after heat treating, it has a yield strength of 350 to 400 ksi. No other metal has such a high room temperature strength, and no other strengthening mechanism has a greater effect than the interstitial alloying of carbon in iron. The details of strengthening mechanisms are discussed in a following section.

Figure 2-22(b) shows an interstitial atom, I, in the solvent matrix. The atom, however, lies not in the plane of the solvent lattice but either above or below the sketched plane. Also, the foreign atom is always larger than the "void" space it occupies, so it necessarily forces the surrounding solvent atoms out of their normal array. Therefore, the crystallographic planes are warped outward in the vicinity of an interstitial defect.

The third type of point defect, the presence of a foreign atom at one of the lattice points, is referred to as a *substitutional defect*. When an alloy is made by adding solute atoms that replace (substitute for) solvent atoms in the lattice structure it is called a substitutional alloy. This type of alloy is the most common one for metals. Figures 2-22(c) and (d) show the substitution of a smaller and a larger atom, S, at one of the lattice points. Unlike the interstitial atom, the substitutional one is in the plane of the solvent matrix. The crystallographic planes are also warped in the vicinity of the substitutional defect; inward for the smaller atom and outward for the larger atom. Understanding the distortion of the crystallographic planes is very important to assessing the strength of materials.

The fourth type of point defect, atoms that are displaced from their normal position, occurs in several forms. The atoms in the "contacting" planes of two adjoining crystals are not in their normal position as a result of the crystals having solidified from the liquid state without being in perfect registry with each other. This is considered to be a *grain boundary* defect; such defects have a significant effect on the strength of all polycrystalline materials.

Two additional types of atom displacement defects, which are not present in metallic bonded crystals, occur in ionic crystals. A vacancy defect in an ionic crystal that is associated with a displaced pair, one cation and one anion, is called a *Scholtky defect*. A *Frenkel defect* occurs when a small cation moves from a lattice point, leaving a vacancy, and occupies an interstitial site.

Line defects or dislocations Examination of crystals under the electron microscope have shown interruptions in the periodicity of the lattice structure in certain directions. In a two-dimensional representation these interruptions appear as lines, hence the name line defects. It is believed that a perfect crystal of a metal such as pure iron should have a strength of one or two million psi, whereas in reality such metals have a yield strength of only a few thousand psi. The reason for the three-orders-of-magnitude difference between the postulated and the actual strength of metal crystals is the presence of these line defects.

The two most common line defects are *edge dislocation* and *screw dislocation*. An edge dislocation is the line defect that results from the presence of an extra plane of atoms in one portion of a crystal compared with the adjacent part. Actually, this defect is the edge of this extra plane of atoms and runs from one end of the crystal to the other end. When one is looking at the crystalline plane that is perpendicular to the dislocation line, the imperfection appears as an extra row of atoms in a part of the crystal. Figure 2-23(a) illustrates an edge dislocation in a crystal.

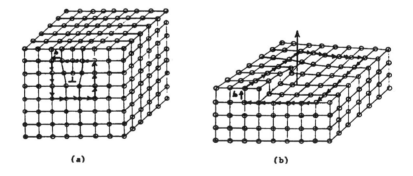

Figure 2-23. Line defects in crystals. (a) Edge dislocation of unit Burgers vector. (b) Screw dislocation of unit Burgers vector.

An edge dislocation is customarily represented by the symbol ⊥, where the vertical leg designates the extra plane of atoms. When the vertical leg is above the horizontal leg, as in Figure 2-23(a), the dislocation is considered positive. When the extra plane of atoms is in the bottom portion of the crystal, the vertical leg is placed below the horizontal one and the dislocation is said to be negative. The part of the crystal containing the extra plane of atoms is in compression while that portion on the other side of the dislocation line is in tension. A dislocation line, since it runs completely across a crystal, deforms the lattice structure to a greater extent than does one point defect.

A screw dislocation is a crystal defect in which the lattice points lie on a spiral or helical surface that revolves around a centerline, which is called the dislocation line. A screw dislocation terminates at a crystal surface, as illustrated in Figure 2-23(b). Shear stresses are set up in the lattice surrounding a screw dislocation because of the distortion in atomic array that the defect causes.

The *Burgers vector* is the distance, measured in multiples of the lattice parameter, that is needed to close a straight-sided loop around a dislocation by going the same number of lattice distances in all four directions. The term is used to define the size of a dislocation and is designated by the letter b. Figure 2-23 shows an edge and a screw dislocation the size of one Burgers vector. A characteristic of an edge dislocation is that it lies perpendicular to its Burgers vector, whereas a screw dislocation lies parallel to its Burgers vector.

Planar defects There are several types of planar (or surface) defects that are caused by a change in the orientation of crystallographic planes across a surface boundary. The most important planar defect is the *grain boundary*, which is the imperfect plane surface that separates two crystals of different orientation in a polycrystalline solid. Grain boundaries originate when the last few remaining atoms of a liquid freeze onto the meeting faces of two adjacent crystals that have grown from the melt, or, similarly, when two adjacent crystals that have grown by recrystallization meet each other.

Figure 2-24(a) is a cross section of a grain boundary between two cubic crystals A and B of the same material having unit cell length a but greatly different crystallographic orientations. Since the sketch of two crystals is two-dimensional, the meeting surface appears as a line. In three dimensions it would be a plane. However, it can be seen from the sketch that the atoms near the central portion of each crystal are in their normal lattice sites. But for a couple of atomic diameters on either side of the imaginary interface the atoms are severely distorted from their normal lattice points, and the crystallographic

58   Chapter 2

planes are warped where they terminate at the grain boundaries. The material in the grain boundary is at a higher energy level than the material near the center of the grain because of the increased elastic strain energy of the atoms that are forced from their normal (lowest energy) sites in a perfect lattice. This higher energy level and lattice distortion cause the grain boundary material to be stronger, have a higher diffusion rate, and serve as a more favorable site for the nucleation of second phases than the interior material.

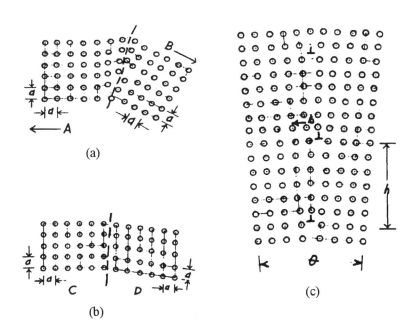

Figure 2-24. Two-dimensional representation of some planar defects. (a) Grain boundary. The large arrows point toward the center of the grains A and B. (b) Twin boundary. The twins C and D constitute one grain with only a slight misalignment of the crystalline planes. (c) Low-angle grain boundary where b/h. Actually h is larger than shown here.

Another important planar defect is the *twin boundary*, which is the plane that separates two portions of a single crystal having slightly different orientations. The two twins are mirror images of each other. For example, in Figure 2-24(b), which is a two-dimensional representation of twins, if the twin

D were folded over about the twin boundary as an axis, its lattice points would correspond to those of twin C. The distortion of the twinned lattice is low in comparison to a grain boundary. Twins which form in most FCC metal crystals, especially the copper- and nickel-base alloys, during freezing from the melt or recrystallization are called "annealing twins." Twins which form in some metals during cold work (plastic deformation) are called "mechanical twins."

A third planar defect is the *low-angle grain boundary* or *low-angle tilt boundary*, where the angular misalignment of the two grains is very small, of the order of a few degrees. In a sense it is a very undistorted grain boundary. The angular mismatch of the crystal planes is due to a row of dislocations piled above each other. Figure 2-24(c) is a two-dimensional representation of a low-angle grain boundary.

A *stacking fault* is a planar defect that occurs when one crystalline plane is stacked out of its normal sequence in the lattice array. The lattice on both sides of the defect is normal. For example, the normal FCC stacking of planes may be interrupted by one layer of an HCP plane since both are close-packed structures with atomic packing factors of 0.74. Such stacking faults can occur during the formation of a crystal or by plastic deformation.

The above brief discussion of crystal dislocations and other defects is sufficient to explain the theories behind each of the strengthening mechanisms and to apply these theories to the control of the mechanical properties of real materials. It is necessary to go further into dislocation theory only if one is to do research in that field. It is much more beneficial for an engineer to understand the control of mechanical properties on the basis of slip, which is the movement of extremely large numbers of atoms along a crystalline plane; for then it is possible to select and specify the optimal condition for any material.

Slip in Crystalline Solids

Slip can be defined as the massive sliding movement of one large body of atoms with respect to the remaining body of atoms of the crystal along crystallographic planes. Slip can also be considered an avalanche of dislocations along one plane that pile up at grain boundaries or inclusions. The planes along which slip occurs are called slip planes. Slip occurs only with relatively high stresses, i.e., greater than the yield strength, and causes plastic deformation.

When a crystalline solid, or a single crystal, is subjected to low loads, the atoms move slightly from their normal lattice sites and return to their proper position when the load is removed. Figure 2-25 illustrates the relative movements of the atoms in a crystal during elastic deformation, where no slip occurs under tensile, compressive, and shear loading. The displacements of the individual atoms are very small during elastic deformation. They are submicroscopic, a fraction of an atomic distance. Although there are some dislocation movements, they are few in number, cover very short distances, and are reversible.

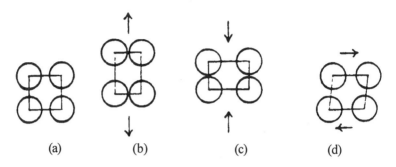

Figure 2-25. Atomic movements in a crystal lattice during elastic deformation: (a) no stress, (b) tensile stress, (c) compressive stress, and (d) shear stress. The atoms return to their original position upon removal of the stress.

Slip, on the other hand, is microscopic in scale and causes plastic (permanent) deformation that is macroscopic. Figure 2-26 contains several two-dimensional lattice arrays, which in a simplified manner illustrate the mechanism by means of which slip takes place. A typical perfect cubic lattice is shown in (a), which is a small part of a single crystal. If sufficiently large shear stresses τ are placed on the crystal, all of the atoms above the labeled slip plane move to the right simultaneously with respect to the atoms below the slip plane, as shown in (b). The lattice is still a perfect cubic structure; only the outline or exterior shape of the single crystal has changed. It is believed, on the basis of the theories of elasticity, that the shear stress must be equal to the value of $G/2\pi$, where G is the shear modulus of elasticity. G is related to Young's modulus of elasticity E and Poisson's ratio by the equation

$$G = \frac{E}{2(1+\mu)} \qquad (2\text{-}3)$$

For iron, $E = 30 \times 10^3$ ksi and $\mu = 0.3$, so $G = 11.5$ ksi. Therefore, the so-called theoretical shear strength for slip to occur in iron is

$$\tau = \frac{G}{2\pi} = \frac{11.5}{2\pi} = 1800 \text{ ksi} \tag{2-4}$$

However, slip occurs in iron crystals with shear stresses of only 4 to 5 ksi, which is more than 2 orders of magnitude smaller. The theoretical shear strength of the other pure metals is also 400 to 500 times larger than the actual shear strength. The commonly accepted explanation of why the actual shear stress is so much lower than the theoretical value is that slip does not occur by the simultaneous movement of all of the atoms along the slip plane but rather by the movement of individual rows (the dislocation row or plane) of atoms. Thus it is the movement of dislocations along the slip plane to the grain boundary that causes the actual shear stress for plastic deformation to be so low. Figure 2-26(c), (d), and (e) illustrates the movement of a dislocation that results in slip.

In real crystals of metals, slip terminates at the grain boundaries or the free surface faces and causes substantial jogs or steps, much larger than shown in Figure 2-26. The spacing of slip planes and the size of the jog has been studied experimentally for some of the common metals. The spacing of the parallel planes along which slip occurs varies randomly, with an average distance between slip planes of about 2000 atom diameters. The length of the step or jog at the surface of the grain is approximately 200 to 700 atom diameters.

The atomic displacements associated with slip, unlike that of the initial movements of dislocations, is irreversible in that the slip jog remains when the shear stresses are removed. That is, slip causes a permanent change in shape, called *plastic deformation.*

The evidence of slip on metallurgically prepared samples is seen as slip lines when examined under a microscope. The slip lines are the intersection of the crystallographic planes along which slip occurred and the etched surface of the specimen. Slip results in a narrow band on either side of the slip plane, within which the lattice structure is severely distorted. These slip lines do not appear on the face of a specimen that is metallurgically polished after slip occurs, but they appear after etching with a suitable chemical reagent that dissolves the metal being studied. The slip lines become visible for the same reason that grain boundaries are visible after etching: the internal energy of the material within the distorted area is considerably higher than that of the material within the rest of the crystal. The metal in the higher energy level dissolves into the reagent much more rapidly than the rest of the crystal, thus leaving a narrow groove where the severely distorted band intersects the surface. Slip lines can also be seen on unetched specimens if the metal is polished prior to being plastically deformed.

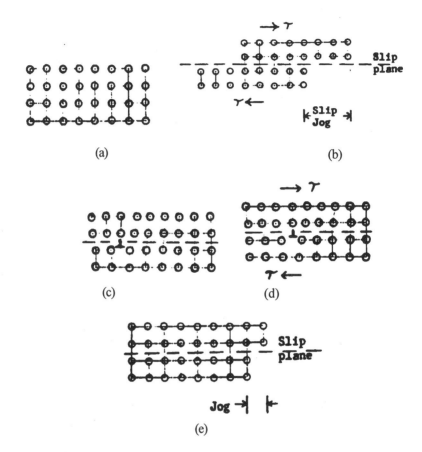

Figure 2-26. Slip mechanism. (a) A perfect crystal. (b) Idealized slip in a perfect crystal. (c) Part of a crystal with one edge dislocation. (d) Movement of dislocation subject to shear stress. (d) Jog produced in the crystal face by dislocation motion.

The importance of understanding the slip mechanism as a prerequisite to studying the mechanical properties of materials can be appreciated by

The Structure of Solids   63

knowing that slip and yield strength (as defined in Chapter 7) are two terms that express the onset of plastic deformation from two different viewpoints: a microscopic and a macroscopic view.

MECHANICAL STRENGTH

Although the specific mechanical properties of real materials are discussed in great detail in Chapter 7, it is appropriate at this time to relate the concepts of the strengthening mechanisms to the previously described crystalline structures. Mechanical properties can best be studied on the basis of three precepts which encompass all of the strengthening mechanisms. These three principles are stated here because they involve the distortion of the lattice structure that has just been discussed. They will be applied in the later chapters dealing with the control and prediction of mechanical properties.

Principles of Mechanical Strength

Strength was previously defined as a material's ability to resist slip. It follows that the first principle of mechanical strength is: *a material is strengthened when slip is made more difficult to initiate.* To make a material stronger it must be given a treatment that retards the avalanche of dislocations or, in other words, "pegs" the slip planes.

The second principle of mechanical strength is: *slip is retarded by inducing mechanical strains, or distortions, in the lattice structure of the material.* These distortions were discussed previously as lattice imperfections or defects. Thus it is a paradox that the source of strength in real, polycrystalline materials is crystal "imperfections" or "defects."

The third principle of mechanical strength is: *there are four methods to induce the mechanical strains or lattice distortions in a material, namely: decreasing the grain size; ulastic deformation, (referred to as cold work); single-phase alloying; and multiple-phase alloying.*

These four methods were discussed previously in the section "Crystal Imperfections" in regard to the manner in which they distort the lattice structure. On that basis they are considered as defects. Now we want to consider them as strengthening factors and to illustrate how they affect the strength of a material.

Grain size. The local distortion of the lattice structure at the grain boundaries induces substantial strain energy in those regions. This distortion impedes slip, or causes the dislocations to pile up, and consequently the grain boundary material is stronger than the material at the central portions of the crystal. This is true for most metals at room temperature. However, as additional energy is added to a polycrystalline material by raising the temperature, the grain-boundary material softens (and also melts) sooner, or at a lower temperature, than the bulk of the grain. At some temperature, called the *equicohesive temperature*, the strengths at these two regions are equal. And above the equicohesive temperature, the grain boundary material is the weaker of the two. This explains why materials that are used at elevated temperatures have higher creep strengths when their grains are coarse rather than fine.

The ratio of surface area to volume of a sphere is inversely proportional to the diameter. Therefore, as the diameter of a sphere decreases, its ratio of surface area to volume increases. This means that for a given weight, or volume, of a polycrystalline solid, the total grain-boundary surface increases as the grain size decreases. Since the grain-boundary material is stronger than the interior material, the strength also varies inversely as the grain size. Also, since the surface area of a sphere is proportional to the square of the diameter it can be assumed as a first approximation that the yield strength is proportional to the reciprocal of the square of the grain diameter. That is, $S_y = kd^{-0.5}$ or $S_y = c + kd^{-0.5}$.

Figure 2-27(a) shows how the 0.2% offset yield strength of 70Cu-30Zn brass varies with grain size. In this case the yield strength increases by a factor of 4 with a grain diameter ratio of 24. The strength of some materials, such as aluminum or steel, is not as greatly affected by grain size.

Cold work Cold work is significantly more beneficial to strengthening than is grain size for most metals. When a crystalline material is plastically deformed, there is a mass movement of a body of atoms along a crystallographic plane (an avalanche of dislocations) that terminates at the grain boundaries. This movement in a polycrystalline material distorts both the grain boundaries, the crystalline planes in the grain in which slip occurs, and the adjacent grains. This is evident in Figure 2-26(b) and (e), where it can be seen that the slip "jogs" actually make pronounced steps at the grain boundaries. Actually, a portion of one grain intrudes into the space that was previously occupied by another grain, with a resulting distortion of the lattice in both grains.

Figure 2-27(b) illustrates the effect of cold work on the yield strength of 70Cu-30Zn brass. With only 10% cold work, the yield strength is raised by a factor of 3.5. And 60% cold work increases the strength nearly eightfold. In general, 10% cold work more than doubles the yield strength of most metals.

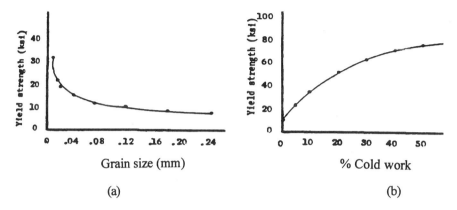

Figure 2-27. Yield strength vs. grain size (a) and percent cold work (b) for 70Cu-30Zn brass.

Single-phase alloying Alloying (single and multiple phase) is the most important of the methods available to control or manipulate the mechanical properties of materials. The greatest increase in strength occurs when iron having a yield strength of 10 to 20 ksi is alloyed with less than 1.0% carbon to form a single-phase (martensite) that has a yield strength of nearly 400 ksi.

The lattice is distorted and dislocation movement is impeded when foreign (solute) atoms are added to the lattice structure of a pure material. Figure 2-22(b), (c) and (d) illustrates this condition, which was discussed in the section "Point Defects." Introducing vacancy defects, as shown in Figure 2-22(a), is not a practical strengthening method. The reason single-phase alloying has such a great effect on strength is that the entire lattice network is distorted uniformly, whereas, in the other mechanism, there are regions in each crystal that are severely distorted and other regions that are hardly distorted at all.

Figure 2-28(a) shows the effect on the strength of the material of adding a foreign element B or C to the lattice structure of element A. From this figure it is clear that the strengthening effect is not the same for all elements. In general, the further the ratio of solute atom diameter to solvent atom diameter is from unity the greater the strengthening effect will be. However, the further this ratio is from unity, as previously explained, the less soluble the two atoms are in each other's lattice.

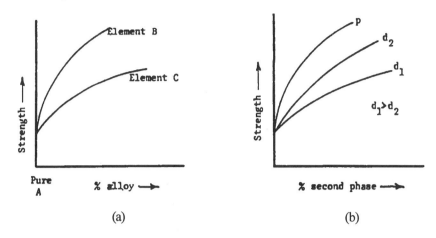

Figure 2-28. The effect of alloying on strength. (a) Single-phase alloying. Atomic diameter ratio B/A > C/A. (b) Multiple-phase alloying. P is a nonspherically shaped particle; d is the spherical particle diameter.

Multiple-phase alloys Multiple-phase alloys can be made in three different ways. One method is by annealing. In this case the alloy is heated to a single-phase region where the second element is completely soluble in the first. On slow cooling the second phase precipitates as a massive network in the grain boundary regions of the solvent matrix. This is the least effective form of achieving strengthening through alloying. The second method is similar except that the alloy is rapidly cooled from the high single-phase region so that a supersaturated solid phase occurs at room temperature. This material is then reheated to a relatively low temperature so that the second phase precipitates throughout the entire crystal as extremely fine particles rather than concentrating at the grain boundaries. This is the common *precipitation hardening* procedure. The third method is to add a compound, in the form of small particles, that is insoluble in the parent material. Thus, the two phases must be mixed in powder form and then sintered. This method is called *dispersion hardening*. At the present time there are only about a half dozen dispersion-hardenable alloys in commercial use. The most notable ones are Al_2O_3 particles in aluminum (called SAP, for sintered aluminum powder) and ThO in nickel.

REFERENCES

1. Barret, C.S. *Structure of Metals,* 1st ed. McGraw-Hill, New York, 1943.
2. Seitz, F. *The Physics of Metals,* 1st ed. McGraw-Hill, New York, 1943.
3. Cottrell, A.H. *The Mechanical Properties of Matter.* John Wiley & Sons, New York, 1964.
4. van Vlack, L.H. *Elements of Materials Science,* Addison-Wesley, MA, 1989.
5. Flinn, R.A. and P.H. Trojan, *Engineering Materials and Their Applications,* Houghton Mifflin, Boston, 1990.
6. Shackelford, J.F. *Material Science for Engineers,* Macmillan Publishing Co., New York, 1992.
7. Murray, G.T. *Introduction to Engineering Materials,* Marcel Dekker, Inc. New York, 1993.

Chapter 3

EQUILIBRIUM PHASE DIAGRAMS

INTRODUCTION

Equilibrium phase diagrams are graphs that indicate which phases exist at any specific combination of temperature, pressure, and composition for a given material. They are one of the most important tools that an engineer has available to determine how a given material can be treated to achieve the best mechanical properties. They are a vital guide to the proper selection and specification of metals. The term "equilibrium" in this context means that the material is held at the higher temperatures for a sufficiently long time for it to change to its most stable condition at that temperature. And any subsequent cooling is done sufficiently slowly for the material to change to its most stable condition at each lower temperature. In practice, furnace cooling is considered to be slow enough to form the equilibrium phase.

Phase diagrams are classified as unary, binary, ternary, etc., on the basis of the number of pure components involved. Representative phase diagrams are shown in Figure 3-1. A unary diagram, consisting of only one component (an element or compound), has temperature and pressure as the two axes and three intersecting lines that divide the graph into three areas; one for each of the three phases, which in this case are the three states of matter (solid, liquid, and gas). The point where the three phases can coexist is called the triple point. Binary phase diagrams are by far the ones most commonly used and they are discussed in detail later in this chapter. They have composition and temperature as the two axes, as shown in Figure 3-1(b). A ternary diagram is made up of three compositions; each side of an equilateral triangle gives the percentage of one of the components as shown in Figure 3-1(c). A vertex of the diagram represents 100% of the component designated at the vertex. The side of the diagram opposite that particular vertex represents 0% of that component. Thus, point P in Figure 3-1(c), which lies approximately 1/3 of the vertical distance between the horizontal base and the apex C, contains 33% of component C.

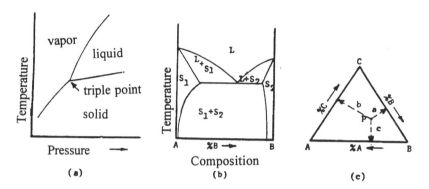

Figure 3-1. Representative equilibrium phase diagrams: (a) unary; (b) binary L = Liquid, S_1 = Solid #1, S_2 = Solid #2; (c) ternary.

PHASES

A *phase* is a chemically homogeneous, physically distinct, and mechanically separable portion of a material system. The pure elemental metal, lead, for example, is a single solid phase at room temperature. Any small sample taken from a polycrystalline piece of lead has the same chemical composition as any other sample. If the piece of metal is heated above its melting temperature (621°F for lead), it again consists of only one phase, but a different phase from the solid one even though they both have the same composition. The liquid phase and the solid phase can coexist at the melting temperature.

If two pounds of tin and eight pounds of lead are placed in a crucible and heated above 621°F, one liquid phase results (see Figure 3-7). When this alloy is cooled back to room temperature it consists of 2 phases. They are physically distinct and mechanically separable on a microscopic scale. One of the solid phases is nearly pure lead and the second solid phase is pure tin, since

Equilibrium Phase Diagrams 71

lead is insoluble in tin at room temperature. These aspects of phases are discussed further under the heading "Binary Phase Diagrams."

A phase may be a pure element, a solution (solid, liquid, or gas), or a compound of two or more elements. Any one of these phases may be the component that designates the end points of a binary phase diagram or the vertex of a ternary diagram. That is, the two components of a binary diagram may be the elements lead and tin, or they may be the compounds Al_2O_3 and SiO_2.

HEATING AND COOLING CURVES

Heat is either absorbed or liberated during a phase transformation, depending upon the phase change involved. The latent heat of vaporization, H_v is associated with the change from liquid to vapor phase; the latent heat of fusion, H_f, is associated with the change from liquid to solid phase; the latent heat of transformation, H_t, is associated with the change from solid phase to one or two of the other phases. These phase transformations are clearly indicated by inflection points or horizontal portions of a heating or cooling curve caused by the evolution or absorption of heat. Most of our equilibrium phase diagrams have been constructed from data obtained by means of heating and cooling curves. The relationship of cooling curves to phase diagrams is important to an understanding of the heat treatments given to the various material systems.

Isothermal Phase Change

If a beaker of mercury with a thermocouple immersed in it, or a piece of lead with a thermocouple attached to it, is placed in a furnace that is held at 600°F, heating curves similar to H in Figure 3-2(a) are obtained if simultaneous time-temperature data are plotted. H is a smooth curve that approaches 600°F asymptotically. If the beaker of mercury, or the piece of lead, is taken out of the furnace and allowed to cool in air, a smooth cooling curve similar to C that is asymptotic to room temperature results. The curves for these two materials in the temperature range of 70° to 600°F are smooth because neither material undergoes any change of state or phase at these temperatures. (Mercury melts at –38°F and boils at 675°F; lead melts at 621°F and boils at 3160°F).

72   Chapter 3

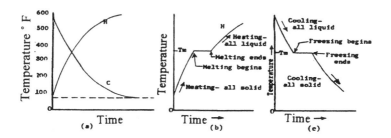

Figure 3-2. Typical heating and cooling curves. (a) No change of state or phase. (b) Heating curve with melting occurring. (c) Cooling curve with freezing occurring.

If a piece of lead with a thermocouple attached to it is placed in a crucible which is then placed in a furnace that is held at 1000°F, a heating curve similar to Figure 3-2(b) results if the temperature-time data is plotted. The curve rises smoothly up to 621°F, which is the melting temperature of lead. At this temperature, the lead starts to melt. The lead absorbs the latent heat of fusion, that is, heat is required to melt the lead — actually 11.3 BTU's per pound. Thus the temperature remains constant at 621°F until all the lead has liquefied because the heat that would have raised the temperature was needed to melt it. The length of the horizontal line indicates the melting time. After all the lead has melted, the temperature again rises smoothly and approaches 1000°F asymptotically. If the crucible is taken out of the furnace and allowed to air cool, its cooling curve will be as illustrated in Figure 3-2(c). The horizontal line at 621°F is the result of the latent heat of fusion being liberated and thus preventing the temperature from dropping until all of the lead has solidified.

Rule 1. All pure elements and chemical compounds have horizontal portions of their heating and cooling curves corresponding to their melting and boiling temperatures.

Some pure elements exist in the solid state in more than one crystal structure, as shown in Table 2-2, depending upon the temperature. This property is called allotropism or polymorphism. For example, solid iron exists in three allotropic forms: α iron, which is BCC, up to 1670°F; γ iron, which is FCC, between 1670 and 2550°F; and δ iron, which is BCC, between 2550 and 2795°F. Although α iron is ferromagnetic up to 1415°F and nonmagnetic at higher temperatures, this distinction is not considered an allotropic change. A cooling curve for pure iron is shown in Figure 3-3. The top horizontal line indicates the change from the liquid to the solid state. The lower two horizontal lines indicate the allotropic transformation from one crystalline solid phase to another crystalline solid phase.

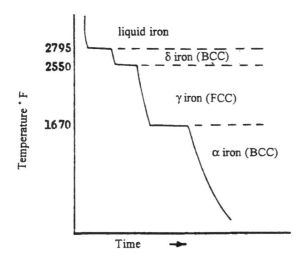

Figure 3-3. Cooling curves illustrating allotropic transformation. The ends of the horizontal line indicate the beginning and the end, respectively, of the transformation.

Rule 2. All pure elements and chemical compounds have horizontal portions of their heating and cooling curves corresponding to their allotropic phase transformation temperatures.

Sometimes a phase transformation occurs when one single phase changes into two different phases. Several examples of this phenomenon are illustrated in the following section on binary phase diagrams. Two of the most common examples of this type of transformation occur during the freezing of a 38% Pb-62% Sn solder alloy at 359°F and the slow cooling at 1333°F of a 0.8% carbon unalloyed steel. The first transformation, a liquid phase changes to two solid phases during cooling and is called a *eutectic* (lowest temperature) reaction. The transformation temperature of 359°F is called the eutectic temperature, and the composition corresponding to the lowest freezing temperature is called the eutectic composition. Figure 3-4(a) shows a cooling curve for the lead-tin eutectic composition. The left end of the horizontal line indicates when the transformation from the liquid state to the solid state begins. However, in the case of eutectics, the "solid state" is really a mechanical mixture of two distinct, separable solid phases: α and β. These two new phases do not have the same composition of the original liquid. Instead, one, α, is rich in lead while the other, β, is nearly pure tin (see Figure 3-7). When solidification begins, the two phases nucleate and grow simultaneously rather than sequentially. The relative amounts of the two phases that are present is discussed in the following section, "Binary Phase Diagrams."

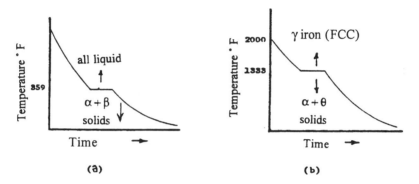

Figure 3-4. Cooling curves illustrating eutectic and eutectoid transformation: (a) 38% Pb-62% Sn lead eutectic, (b) 0.8% C eutectoid steel.

When a plain carbon steel having 0.8% carbon is cooled from a temperature above 1333°F, but lower than the solidus temperature of 2500°F, a cooling curve such as shown in Figure 3-4(b) results. Upon cooling to 1333°F the iron, which is called *austenite* and has an FCC lattice, begins to transform to two different solid phases. One of the phases is an α iron which is called *ferrite* and has a BCC lattice; Fe₃C, the other phase, is the intermetallic compound, identified with the symbol θ, which is called *cementite* or iron carbide and has an orthorhombic structure. This type of transformation, from one solid phase to two different solid phases upon cooling, is called a *eutectoid* reaction. The temperature at which this occurs (it is the lowest transformation temperature) is the eutectoid temperature, and the composition associated with this transformation is called the eutectoid composition.

Equilibrium Phase Diagrams 75

There are two other transformation processes that are less frequently encountered in the study of mechanical properties of materials. A *peritectic* reaction is one in which a solid phase and a liquid phase transform on cooling to a solid phase, different from the original one. An example is the transformation of δ iron and liquid iron at 2718°F to γ iron. A *peritectoid* reaction is one in which two solid phases transform during cooling to a third solid phase. All four of the transformations described above are thermally reversible. That is, upon heating, the opposite reaction occurs and the material returns to the same condition it was in prior to cooling. These four transformations are also called invariant reactions.

Rule 3. All materials of either eutectic, eutectoid, peritectic, or peritectoid composition have horizontal portions on their heating and cooling curves corresponding to the temperature at which that particular reaction occurs.

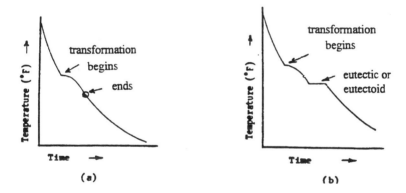

Figure 3-5. Cooling curve showing nonisothermal phase change. (a) Two inflection points. The circle identifies the lower inflection point. (b) One inflection and a short horizontal.

Non-Isothermal Phase Change

The cooling curves for most alloys, except for the few types of compositions described above, have either two inflection points or else one inflection point and a horizontal portion. Figure 3-5 illustrates these two types of cooling curves.

76   Chapter 3

The first type of curve occurs when a liquid phase solidifies over a range of temperatures to a solid solution whose composition changes as the temperature drops. The higher inflection point signifies the beginning of solidification and is very clearly defined. The lower inflection point, which designates the end of solidification, is not well defined since it is the point tangent to two smooth curves. For this type of material, the upper inflection point defines the location of liquidus line on the phase diagram and the lower point determines the location of the solidus line. These are illustrated in Figure 3-6. This type of curve also occurs when one solid solution transforms to a second solid solution over a range of temperatures.

The second type of nonisothermal phase change, as shown in Figure 3-5(b), occurs when a material freezes to a partial eutectic or else transforms to a partial eutectoid, with the balance being another phase. The above aspects of cooling curves (and heating curves, which are similar) will become clearer when they are discussed in conjunction with binary phase diagrams in the following sections.

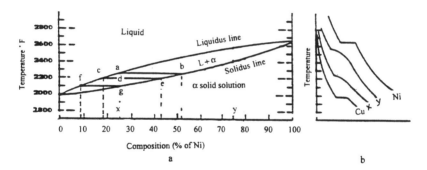

Figure 3-6. Copper-nickel phase diagrams, (a) Complete liquid and solid solubility. (b) Cooling curves for pure copper and nickel plus two alloys.

BINARY PHASE DIAGRAMS

A binary phase diagram is a graph of a two component material system that shows the following for any combination of temperatures and original composition: (1) the phases that are present; (2) the composition of each phase; (3) the amount of each phase present. Because the diagrams indicate what

Equilibrium Phase Diagrams

temperature a material must be heated to in order to undergo a phase change, they serve as a general guide to the selection of materials and their heat treatments. Some specific types of the most common phase diagrams will be briefly discussed so that the reader will know how to use them. Volume 8 of the Metals Handbook (8th Edition, American Society for Metals) is the most complete source of phase diagrams.

Through thermodynamic studies, Williard Gibbs determined the relationship between the number of degrees of freedom (F) — which is the number of variables such as temperature, pressure, or composition that can be changed independently without affecting the number of phases — and the number of components (C) of the materials system. This relationship is known as *Gibbs phase rule* and is expressed as

$$P + F = C + 2$$

Our use of phase diagrams for selecting and understanding materials generally assumes atmospheric pressure, so one degree of freedom is thus specified, and the phase rule reduces to

$$P + F = C + 1$$

The number of components (C) in a binary phase diagram is two, and so, for this special case, the relationship between the number of phases and the degrees of freedom (excluding pressure) is

$$P + F = 3$$

Thus, in a one-phase region of a binary phase diagram, there are two degrees of freedom: the temperature and the composition of the phase. In a two-phase region of a binary phase diagram, there is only one degree of freedom. Thus, if the temperature is specified, the composition of the two phases is fixed.

The simplest type of binary phase diagram is one in which the two components are completely soluble in each other in both the liquid and the solid states. Only a few metal systems, and no complex compound systems, have this type of diagram. Complete solubility in the solid state requires that the atoms of the two components have similar valence electrons, atomic diameters that differ by less than 15 percent, and the same type of lattice structure. Alloys of this type are called substitutional alloys because the atoms of one element

78 Chapter 3

substitute at the lattice points for the atoms of the other element, as described in Chapter 2.

Figure 3-6(a) is the equilibrium phase of the copper-nickel system that has complete solid solubility. All compositions are in the liquid state at temperatures above the liquidus line and in the solid state at temperatures below the solidus line. Both liquid and solid phases coexist in the area between the liquidus and solidus lines. The cooling curves for pure copper and pure nickel in Figure 3-6(b) have horizontal positions at 1981°F and 2650°F, respectively, corresponding to their melting temperatures. Cooling curves x and y are for a 25% Ni and a 75% Ni alloy, respectively. Each curve has two inflection points, the higher one corresponding to the liquidus temperature and the lower one to the solidus temperature.

Three Application Rules

Binary phase diagrams can be most easily interpreted and used by understanding the following three rules. The rules are explained in conjunction with Figure 3-6(a) since the diagram itself is easy to understand.

Rule 1. **(Identity of Phases).** For any combination of temperature and original composition, the phases are defined by the area of the phase diagram in which the point of intersection of the horizontal temperature line and vertical composition line falls.

For example, consider the 25% Ni-75% Cu alloy identified as x in Figure 3-6(a). At 2400°F the temperature and original composition lines intersect in the one phase (liquid) area. When this alloy is cooled to about 2250°F, which is the liquidus temperature for this composition as indicated by point *a*, some solid starts to nucleate. This is the beginning of the two phase region. When the alloy cools to about 2200°F, the intersection of the temperature composition line is *d* which is in the two phase L+α region. Upon cooling to 2100°F the alloy is at the solidus temperature, point *g*, and all of the liquid has solidified. One solid phase exists at all lower temperatures.

Rule 2. **(Composition of Phases).** The composition of a phase in a one-phase region is the same as the original composition of the material. The composition of the phases in a two-phase region is the same as the composition corresponding to the intersection of the horizontal temperature line with the liquidus and solidus lines (or solubility limit lines in the solid state).

Equilibrium Phase Diagrams 79

For example, consider the material composition x at 2200°F. The intersection of these two lines is point *d,* which is in the two-phase region. The horizontal line drawn through *d*, sometimes called a tie line, intersects the liquidus line at *c* and the solidus line at *e*. The composition of the equilibrium liquid phase at 2200°F is determined by projecting the point *c* vertically to the composition axis, which in this case is 18% Ni-82% Cu. Likewise the composition of the equilibrium solid phase at 2200°F is determined from the location of *e* as 43% Ni-57% Cu. All of the solid phases are identified as solid solutions. Thus the composition of α can vary from 0% Ni to 100% Ni, depending upon the original composition of the alloy.

As stated earlier, during equilibrium cooling, all of the solid at a given temperature is of the same composition, regardless of what its composition was when it first solidified. Consider the solid that nucleates and grows from a melt of composition x. At 2250°F the first solid nucleates as a few small grains of composition *b*, 52% of Ni. During slow cooling from 2250°F to 2200°F, the small grains grow larger by more liquid solidifying onto them and the composition of all the grains, including the original center, gradually changes to *e*, 43% of Ni. During the slow cooling, the nickel diffuses rapidly from the center of the solid grains to make the entire grain homogeneous.

At 2100°F the last drops of liquid that contain only 9% Ni solidify onto the surface of the previously formed grains. Since diffusion occurs rapidly at this high temperature, and also with equilibrium cooling, all portions of every grain are of the original composition x, 25% Ni-75% Cu.

During commercial foundry practice in making castings, as well as during fusion welding, the cooling rates are much faster than the equilibrium rate. This has two effects on the material. First, the temperatures at which the transformations occur are lowered. That is, both the liquidus and the solidus lines are lowered. Second, homogenization does not occur and a *cored* structure, having a different composition at the center (core) from that at the outer surface of each grain, results.

The composition of the phases in a two-solid phase region is determined in exactly the same manner as described above.

Rule 3. **(Amount of phases).** The amount of each phase present in a two-phase region is the ratio of the distance between the original composition and the end of the tie line opposite the phase to the distance between the two ends of the tie line.

This rule is known as the *inverse lever rule*. The following example explains the application of this rule.

80   Chapter 3

The tie line for a material of composition x at 2200°F is ce, as explained above. The distance between the original composition, point d, and the end of tie line opposite the solid phase is c. The distance between the two ends of the tie line is ec. Thus, from the inverse lever rule:

$$\text{the amount (fraction) of } \alpha = \frac{\text{distance } dc}{\text{distance } ec}$$

$$\text{or simply } \alpha = \frac{dc}{ec}$$

The numerical value for the amount of α present is calculated by using the composition scale at the bottom of the phase diagram to determine the relative lengths of the lines. Thus,

$$\alpha = \frac{25-18}{43-18} = \frac{7}{25} \text{ or } 28\%$$

In similar manner the amount of liquid is determined as:

$$L = \frac{ed}{ec} = \frac{43-25}{43-18} = \frac{18}{25} \text{ or } 72\%$$

The amounts of material in each phase are calculated on a percentage weight basis. Thus 100 pounds of an alloy of original composition 25% Ni-75% Cu held at a temperature of 2200°F will consist of 28 pounds of a solid solution and 72 pounds of a liquid solution. The composition of the solid solution is 43% Ni-57% Cu, and that of the liquid solution is 18% Ni-82% Cu. The original 25 pounds of nickel in alloy x is distributed, at 2200°F, as follows:

Ni in the solid	= 28 lb. × 43%	= 12 lb.
Ni in the liquid	= 72 lb. × 18%	= 13 lb.
	Total	= 25 lb.

Calculations of this type are referred to as a mass balance and serve as a practical check on the composition of material systems.

Some Common Phase Diagrams

Examples of the most common types of phase diagrams, such as those having eutectics or eutectoids or those of age-hardenable alloys, are discussed here. They will also be used in the later chapters to predict microstructures and mechanical properties.

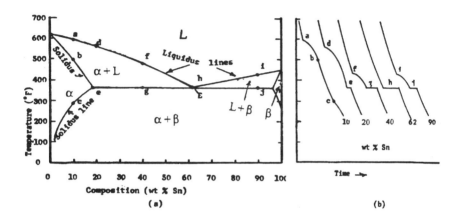

Figure 3-7. Lead-tin phase diagram and related cooling curves.

The lead-tin diagram. The Pb-Sn diagram shown in Figure 3-7(a) is typical of a phase diagram having a eutectic (E in the sketch) as well as limited solid solubility. As in the previous diagram, the liquidus and solidus lines define the composition of the liquid and solid phases, respectively, in the appropriate two-phase region. The composition of the two solid phases in the lower two-phase region is defined by the solidus lines. The solidus lines determine the maximum solubility of the solute atoms in the solvent crystal lattice. Thus, the maximum solubility of tin in lead, which occurs at 359°F (the eutectic temperature), is 19%. The solubility of tin in lead at 100°F is only about 1.0%, and the solubility of lead in tin at 100°F is 0%. The composition of the eutectic is 62% Sn-38% Pb.

The cooling curves for pure lead, which melts at 621°F, and pure tin, which melts at 450°F, are not shown. Consider the cooling curve and associated phase transformations that occur with a lead alloy containing 10% tin.

Upon cooling to 590°F, the liquidus temperature, the first solid, containing approximately 2%, tin appears. The associated evolution of heat results in the sharp upper inflection point a. When the material reaches the solidus temperature of 500°F after slow cooling, it consists of all solid with a uniform composition of 10% Sn-90% Pb. No changes occur during cooling through the α single-phase region. When the temperature reaches 290°F, point c on the solidus line, some pure tin (β phase) begins to nucleate (precipitate) in the grain boundaries of the α crystals. The nucleation occurs in the grain boundaries because of their higher energy level, which is due to the lattice distortion, as discussed in the previous chapter. This precipitation also evolves a small amount of heat causing a third inflection point, less noticeable than the first. Upon further cooling to 100°F more tin (β phase) precipitates, forming small crystals of β between the larger parent grains of α. The composition of the α changes from 10% Sn at 290°F to 2% Sn at 100°F. The amount of the precipitated β that is present at 100°F can be calculated by means of the inverse lever rule. Since the ends of the tie line in this case are 2% Sn and 100% Sn, the amount of β is calculated as follows:

$$\beta = \frac{10-2}{100-2} = \frac{8}{98} \text{ or } 8.2\%$$

The balance of the material, 91.8%, must be α.

Consider next the cooling of the eutectic alloy. It freezes at the constant temperature of 359°F to a mixture of grains of α solid solution containing 19% Sn and grains of β solid solution of 97% Sn. The amount of each phase present, as determined by means of the inverse lever rule, is:

$$\beta = \frac{62-19}{97-19} = \frac{43}{78} \text{ or } 55\%; \text{ and } \alpha = 45\%.$$

When the eutectic cools to 100°F the composition of both phases change: α to 2% Sn and β to 100% Sn. The amount of the two phases also changes.

$$\text{There is now } \frac{62-2}{100-2} = \frac{60}{98} \text{ or } 61\% \text{ } \beta; \text{ and } 39\% \text{ } \alpha$$

As a final example, consider an alloy such as 20% Sn-80% Pb, whose cooling curve has both an inflection point and a horizontal portion. The first solid nucleates from the melt at about 560°F, point d. On cooling through the $\alpha + L$ region, the amount and composition of the two phases change. At 360°F (1° above the eutectoid temperature), the material system consists of 2.3% liquid of 62% Sn-38% Pb composition and 97.7% α of 19% Sn-81% Pb. The final 2.3% of liquid changes at 359°F to a mechanically separable mixture of the two phases α and β. The amount of β present at 358°F is only 1.3%; and at 100°F it is 18.4%. One can test one's ability in using the inverse lever rule by checking these last two values.

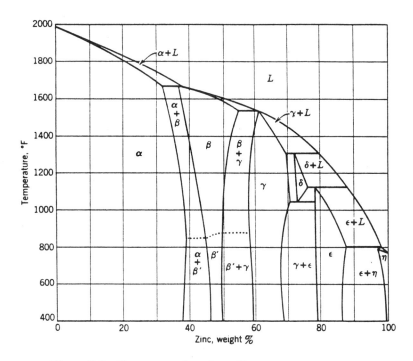

Figure 3-8. The copper-zinc phase diagram.

The copper-zinc diagram. The Cu-Zn phase diagram (Figure 3-8), is very important because it includes all of the brasses. It is typical of a phase diagram in that it is more complicated than the ones previously discussed. It shows an order-disorder reaction at about 850°F, a peritectoid at about 40% Zn, and a eutectoid at about 75% Zn. Most of the brasses contain 40% or less of zinc; the so-called alpha brasses contain less than 35% zinc. The alpha brasses have no allotropic transformations nor other phase changes below the melting range, so they do not respond to heat treating other than stress-relieving or recrystallization after cold work.

An interesting and important characteristic of binary phase diagrams is the 1-2-1 rule. That is, along any horizontal (constant temperature) line from one end of the phase diagram to the other, there is alternately a one-phase region, then a two-phase region.

The aluminum-copper diagram. This diagram can serve as a basis for the study of the 2000 series of aluminum alloys as well as the theory of precipitation hardening in general. The two requirements for successful precipitation hardening of a material system are characteristics of this phase diagram. The first requirement is that there be a single phase at a high temperature in conjunction with a decrease in solubility as the temperature is lowered. Figure 3-9 shows that 5.65% copper can dissolve in solid aluminum at 1018°F, but less than 1/2% is soluble at room temperature. This first characteristic was also true of the lead-tin diagram, which shows that lead-tin alloys do not respond significantly to precipitation hardening. The second requirement is that the second phase that precipitates be a metallic or intermetallic compound. This is true for the Al-Cu system since the second phase, θ, is the compound $CuAl_2$, whereas in the Pb-Sn system the second phase is not a compound — it is pure tin, or β. In this book, as is true in many metallurgy books, the symbol θ is used to denote a compound and the symbols θ_1, θ_2, etc. are used if more than one compound is present. The symbols $\alpha, \beta, \gamma, \delta$, etc., are used to designate solid solution phases. There are two characteristics of a compound in terms of its appearance on a phase diagram. First, it freezes at a constant temperature, and second, it has a narrow range of solubility, usually appearing as a vertical line on a diagram.

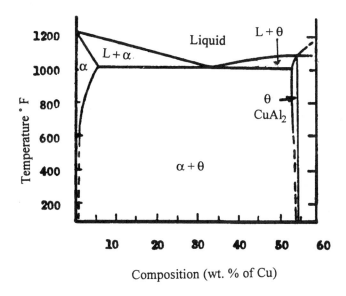

Figure 3-9. The aluminum-copper phase diagram.

The common age-hardenable aluminum-copper alloys contain 5% or less copper. By heating them to 800-900°F, all of the copper is substitutionally dissolved in the solid α aluminum FCC structure. The age-hardening heat treatment, as described in Chapter 5, is a process wherein the second-phase θ is finely dispersed throughout the matrix as opposed to being a massive network around the α grains, as occurs during equilibrium cooling.

The iron-carbon diagram. This is by far the most important phase diagram to the engineer who works with materials, especially metals. It serves as a basis for any thorough understanding of steel. Although it is called the Fe—C diagram, the one most frequently used is really a Fe−Fe$_3$C diagram as illustrated in Figure 3-10. Here the Fe$_3$C(θ) phase is plotted rather than the thermodynamic equilibrium phase, graphite.

The eutectic composition is 4.3% C and the eutectic temperature is 2070°F. There are no useful steels with such a high carbon content. Most steels have a carbon content between 0.08 and 1.2%. Cast irons have a carbon content from 2 to 4%, but they also have 2 to 3% silicon, which alters the phase diagram considerably. The eutectoid composition is 0.8%C and the eutectoid temperature, more frequently called the lower critical temperature, is 1333°F. This diagram is used extensively in the following chapters, which discuss microstructures and mechanical properties.

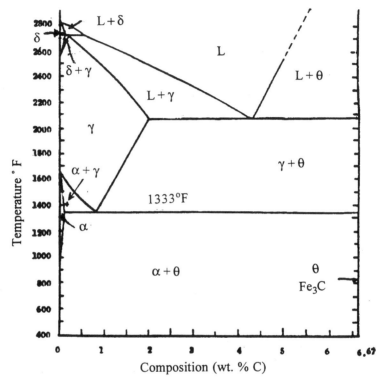

Figure 3-10. The iron-carbon phase diagram. α = ferrite or alpha iron (BCC); γ = austenite or gamma iron (FCC); δ = delta iron (BCC); θ = cementite or iron carbide Fe_3C (orthorhombic). Solubility of carbon in α iron is 0.02% at 1333°F and 0.008% at 70°F. Solubility of carbon in α iron is 2.0% at 2070°F and 0.8% at 1333°F.

The copper-tin diagram. In comparing the copper-zinc phase diagram in Figure 3-8 to the copper-tin diagram in Figure 3-11, it can be seen that copper can dissolve up to 38% zinc and about 16% tin at 1000°F, but, in the latter case, the solubility is nearly zero at room temperature. When more than the equilibrium amount of zinc is added, the second phase that occurs is a solid solution (β) with ductile metallic bond. However, when more than a few percent of tin is added to copper, the second phase that occurs is an intermetallic compound θ having very strong ionic or covalent bonds.

One of the requirements of beneficial precipitation hardening is that the second phase that forms must be a compound. It follows that copper-tin alloys can be significantly strengthened by heat treatment whereas the copper-zinc alloys cannot.

In addition, one of the prerequisites for a material to have good wear resistance is that it contains a dispersion of hard particles in a metal matrix – the harder the particles and the harder the matrix, the better the wear resistance.

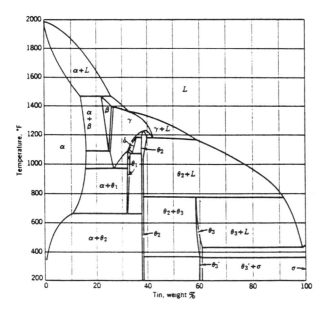

Figure 3-11. Copper-tin equilibrium phase diagram.

The iron-lead diagram. In some cases there is no mutual solubility of two metals at any temperature. Two common cases are iron-silver and iron-lead. The latter case is shown in Figure 3-12. Since lead is not soluble in iron, the lead that is present in leaded steel is there as a second phase and microconstituent that greatly reduces the transverse ductility of the steel.

88 Chapter 3

> (Iron and lead are mutually insoluble in both the liquid and solid state.)

Figure 3-12. Iron-lead equilibrium phase diagram.

The silver solders used in welding steel contain a high percentage of silver. The silver, which will not bond to iron, is added simply to lower the melting temperature of the solder.

REFERENCES

1. Brick, R.M., R.B. Gordon and A. Phillips, *Structure and Properties of Alloys,* McGraw-Hill, New York, 1965.
2. Samans, C.-H. *Metallic Materials in Engineering,* Macmillan, New York, 1963.
3. Murray, G.T. *Introduction to Engineering Materials,* Marcel Dekker, New York, 1993.
4. van Vlack, L.H., *Elements of Materials Science,* 2nd ed. Addison-Wesley, Reading, MA, 1989.
5. Flinn, R.A. and P.H. Trojan, *Engineering Materials and Their Applications,* Houghton Mifflin, Boston, 1990.
6. Shackelford, J.F. *Material Science for Engineers,* Macmillan, New York, 1992.

Chapter 4

MICROCONSTITUENT DIAGRAMS AND MICROSTRUCTURES

INTRODUCTION

The preceding chapters on crystalline structures and material phases form a background to the concepts of microconstituent diagrams and microstructures presented in this chapter. The subject of microstructures is very important because it serves as a means to analyze and compare the mechanical properties of all materials. The common alternative, as is frequently done, is simply to list separately for each family of materials the common treatments and resulting properties. That approach yields tables listing the properties of the various steels, and other tables for the aluminum alloys, and still others for the titanium alloys, but it cannot reveal the fundamental ties among each family of materials. When the treatments and properties of each type of material are discussed separately, the differences among them are emphasized. When the treatments and properties of materials are studied on the basis of microstructure, however, the similarities among them are emphasized. One big advantage of this latter approach is that instead of being forced to select a material on the basis of its properties for a few standard treatments, one can specify a treatment that will give the material optimal properties for a given part and application.

MICROSTRUCTURES

Microstructure is the geometric configuration or grain structure of a polished and etched plane surface of a material as seen through an optical microscope. The usual magnification is $50\times$ to $1500\times$. Thus microstructure is an intermediate-size structure between the macro (with no magnification) scale and the atomic scale. The individual grains that make up the microstructure are called *microconstituents*.

Although the terms phase, microstructure, and microconstituents are similar and related, they do not all mean the same thing. The term *phase* is used to identify a portion of a material system on the basis of chemical composition. *Microconstituent* is used to identify a portion of a material system on the basis of its visual appearance, after polishing and etching, under an optical microscope. *Microstructure* is used in describing the pattern or arrangement of many grains of one or more microconstituents. The following paragraphs discuss these terms and apply them to a sufficient number of material situations so that the reader will soon become thoroughly familiar with them.

MICROSCOPY

In order to understand and interpret microstructures, it is first necessary to know what is being observed and why the structure can be seen. A special type of microscope (a metallurgical microscope, which employs a beam of light rays that go down the barrel of the microscope, are then reflected off the specimen, and go back up the microscope barrel to the eyepiece) must be used to study the granular structure of the materials. If a smooth, plane surface is placed perpendicular to the microscope axis, all of the light rays are reflected back up through the eyepiece and the surface appears uniformly bright or "white." The surface is completely featureless except for any polishing scratches or insoluble inclusions (referred to as "dirt") that are present. The boundaries between the surface layer of grains cannot be seen on a ground and polished specimen. The grain boundaries become visible when the polished specimen, which may be metallic or nonmetallic, is etched with a chemical solution that dissolves the particular material which is being examined. The grain boundaries are seen as a network of dark lines surrounding the white grains. These dark lines are present because the grain boundary dissolves more rapidly in the etchant due to its higher energy level. This leaves the grain boundary in the form of a valley or sharp depression.

Microconstituent Diagrams and Microstructures   93

In sketching microstructures on the basis of microconstituent diagrams, it is expedient to draw a circle to represent the field seen through the eyepiece and to divide the circle into two or more parts with straight lines to represent the grain boundaries. Straight lines are used because they are easier to draw. The sketches shown in Figure 4-1(b) and 4-2(b) take considerable time, effort, and imagination to draw. Since the grains of a real sample are of different size and shape, it is not worth the effort to try to make an exact copy of the microstructure as seen at one small location. The intent is to try to relate to the viewer the significant features of the microstructure such as size of the grains, relative amounts of each microconstituent, and the identity of each. Inclusions and plastic deformation, if present, should also be shown.

Figure 4-1(a-1) helps make the above discussion clearer by showing a side view of a section of a polished specimen. The five grains, A to E, have their tops cut off to form the plane surface. A thin layer of the material, if it is metallic or ductile is severely plastically deformed. All of the light from the microscope is reflected back up through the eyepiece and no microstructure can be seen, as shown by the uniform white area within the circle of Figure 4-1(b-1).

When the polished surface of the specimen is immersed in the proper chemical etchant for a few seconds the "flowed" surface layer is dissolved away and also some of the grain boundary material, as shown in Figure 4-1(a-2). The light rays that impinge on the grains are reflected upward through the eyepiece, but those rays that fall in the grain boundary grooves are reflected off at an angle and therefore appear black to the observer Figure 4-1(b-2) shows the appearance of this network when viewed through a microscope.

If the specimen is etched for a longer time, say 30 seconds or more, then enough of the flat face of some of the grains is dissolved to make the surface rough. The surface of the individual grains would be preferentially attacked depending upon the crystallographic orientation of the grain with respect to the flat surface. This is shown in Figure 4-1(a-3) where the surface of grain A consists of many sharp, shallow grooves. All of the light striking this grain is reflected to the side, and the entire grain appears black. The surface of grain B remains fairly flat and smooth, and it still reflects nearly all of the light striking it. The surface of grain C has wide, shallow grooves and still reflects a little light, making that grain appear "gray." Figure 4-1(b-3) shows the appearance of the material in that condition.

In the preceding example, all of the grains are of the same composition and the same single phase. They may be a pure metal such as copper or a compound such as iron carbide. The geometric structure as it appears in Figure 4-1(b-2) and (b-3) is called the microstructure. Since all the grains in this case are the same phase, the microstructures shown in Figure 4-1(b-2) and (b-3) have only one microconstituent, and in this case it is the same as the phase.

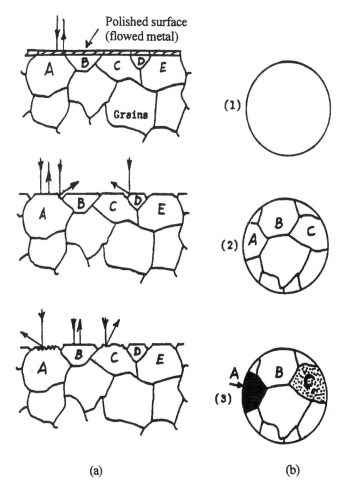

Figure 4-1. Schematic of a specimen under microscopic study: (a) side view of a section with the plane surface at top; (b) top view or appearance as seen through the eyepiece; (1) after grinding and polishing, (2) after a light (short duration) etch, (3) after a deep (long duration) etch.

Microconstituent Diagrams and Microstructures

To distinguish among these three terms, consider Figure 4-2 which shows a material system that consists of approximately half solid solution and half eutectic (or eutectoid). This could be either the 40% Sn-60% Pb alloy or a 0.40% C unalloyed steel, both of which were discussed in the previous chapter on the basis of phases. Grain B is all one phase, α. Grains A and C are made up of alternate platelets α and β, (θ, Fe_3C if it were the steel). Although the α and β platelets are actually small grains, it is the total of all the parallel platelets in a given area, such as A or C, that is considered to be the grain. The appearance of the eutectic grain from a side view is as shown in Figure 4-2(a), where the β platelets have not been etched as much as the α. The appearance, as seen through a microscope, is illustrated in Fugure 4-2(b) and is referred to as the *microstructure*. Now the microconstituents are the grains of α and eutectic (or eutectoid). Thus the entire grain A is one microconstituent, eutectic. C is another grain of the same microconstituent. Grains of α solid solution, such as B, make up the rest of the microconstituents. Thus there are two microconstituents in this microstructure, α and eutectic. There are also two phases in this material system, α and β. The α is both a phase and microconstituent, but the β is only a phase. The platelets in the eutectic are the same composition as the α in the microconstituent grains. This is discussed further in the following section.

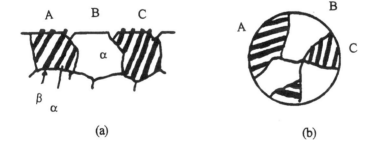

Figure 4-2. Microstructure of a eutectic plus a solid solution.
(a) Side view showing the α phase etched more than the β.
(b) Microscopic view. This geometric structure is called the microstructure.

MICROCONSTITUENT DIAGRAMS

The selection and specification of the condition that a material should be in to achieve its optimal properties for a given application is referred to, more briefly, as the control of mechanical properties. This control is best exercised on the basis of the material's microstructures and microconstituents. Even though the term microstructure refers to the granular structure of the material as seen through a microscope, a microscopic examination is not the best way to study a material's microstructure. The best way to determine a material's microstructure after a given heat treatment is to study its equilibrium phase diagram and non-equilibrium transformation diagrams. This determination can then be confirmed by microscopic examination.

In order to study microstructures this way, the equilibrium phase diagram must be altered to a *microconstituent diagram*. This is a graph of a two-component material system that shows, for any combination of temperature and original composition, what microconstituents are present as well as how much of each is present. To change a phase diagram into a microconstituent diagram, it is necessary simply to draw vertical lines down to the composition axis from eutectic or eutectoid points, and to add these two mechanical mixtures of phases to the diagram. Visualizing this alteration allows one to use a phase diagram as a microconstituent diagram. The inverse lever rule is applied to the microconstituent diagram to determine the amount and composition of the microconstituents in the same manner that it is applied to a phase diagram to determine the amount and composition of the phases.

The Lead-Tin Microconstituent Diagram

Figure 4-3(a) is the Pb-Sn phase diagram of Figure 3-7, modified to make the Pb-Sn microconstituent diagram. To perform this alteration, the vertical dashed line at the eutectic composition was drawn to divide the single $\alpha + \beta$ two-phase region of Figure 3-7 into a pair of two microconstituent regions. The left one is labeled $\alpha + E$ and the right one $E + \beta$ Thus, there are three microconstituents (α, β, and E) in the original two-phase $\alpha + \beta$ region. The following examples demonstrate the use of the microconstituent diagram to determine the microstructure of a material.

Figure 4-3(b) illustrates how the microstructure of a 19% Sn-81% Pb alloy would appear at a temperature of 358°F, 1 degree below the eutectic temperature.

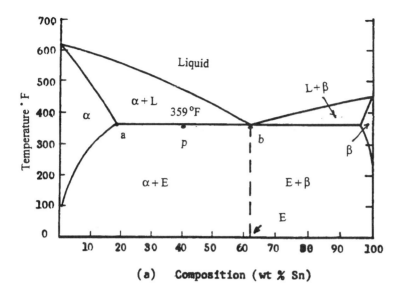

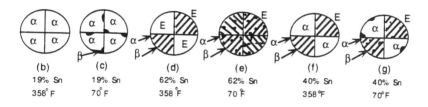

Figure 4-3. The Pb-Sn microconstituent diagram and representative microstructures.

The circles in microstructure sketches are divided into two parts for coarse grains, eight parts for fine grains to indicate the approximate grain size. Figure 4-3(b), with four grains, indicates average or moderate size. To indicate one microconstituent, all the grains are labeled with the same Greek letter. A grain of eutectic (or eutectoid) is designated either by drawing parallel lines in the grain or else by labeling it E, as in Figure 4-3(d). Fine parallel lines, closely spaced, indicate fine platelets or lamellae of the two phases. Broad parallel lines indicate a coarse lamellar structure. Cold work is implied by drawing all of the grain boundaries parallel to each other. A microstructure consisting of 25% α and 75% β would be indicated by four grains, one labeled α and three labeled β.

At 358°F the 19% Sn alloy as shown in (b) consists of one microconstituent, which is the phase α solid solution. As the alloy cools to room temperature slowly, some β precipitates in the boundaries of the α grains. The room temperature microstructure appears as sketched in (c). The matrix is still the α microconstituent, but it now contains only 1% Sn substitutionally dissolved in the lead lattice. The β microconstituent, pure tin, appears as extremely small grains in the boundaries of the larger α grains. In this case, α and β are also the phases.

An alloy of eutectic composition (62% Sn-38% Pb) at 358°F has a microstructure as sketched in (d). The entire microstructure consists of only one microconstituent, eutectic. Two ways of identifying the eutectic are shown: two grains are sectioned with alternate light and dark lines to represent the α and β phases; two grains are simply labeled E. After slowly cooling to room temperature the microstructure appears as sketched in (e), which is identical to (d) except for the additional β that precipitates out of the eutectic α. Since the amount of α that precipitates out of the β is so small, it is neglected.

Now consider an alloy of 40% Sn at 358°F. These two coordinates locate a point p in the two-microconstituent region. The tie line is ab in this case. By means of the inverse lever rule, the amounts of the microconstituents are calculated as follows:

$$E = \frac{pa}{ba} = \frac{40-19}{62-19} = \frac{21}{43} \text{ or } 49\%$$

$$\alpha = \frac{bp}{ba} = \frac{62-40}{62-19} = \frac{22}{43} \text{ or } 51\%$$

or approximately equal amounts of each, as shown in Figure 4-3(f). The α grains have a composition of 19% Sn-81% Pb and the eutectic grains have an overall (average) composition of 62% Sn-38% Pb. After a slow cool to room temperature, the microstructure appears as in (g), where some additional β has precipitated out of both the proeutectic α grains as well as the eutectic α platelets. The amount of eutectic present as a microconstituent at room

Microconstituent Diagrams and Microstructures

temperature is the same as the amount present at the eutectic temperature. It is only the composition of the phases that make up the eutectic that changes.

The Iron-Carbon Microconstituent Diagram

Since the steels are the most widely used material for structural components, the iron-carbon microconstituent diagram is discussed here, and some examples are given to demonstrate the common microstructures that are associated with it. Figure 4-4(a), which depicts the Fe-C microconstituent diagram, was made by adding the two dashed vertical lines to the phase diagram of Figure 3-10 and relabeling the 2 two-phase solid regions. The line from the eutectic (called *ledeburite*) is not drawn below the eutectoid temperature because the eutectic does not exist below the critical temperature. The eutectoid in the Fe-C system is called *pearlite*, a mechanical mixture of α iron and cementite, and is labeled P. The α and θ phases in pearlite are lamellar except after prolonged heating just below the critical temperature, in which case the platelets of θ become small spheroids and the material is then called *spheroidized pearlite*.

The microstructure of commercially pure α iron (less than 0.02 C) is illustrated schematically in Figure 4-4(b). Since the amount of carbon that precipitates out on cooling from 1333°F to 70°F is so small, the microstructure is the same at both temperatures and consists of the one microconstituent, α.

The microstructure of a slowly cooled 0.8% C-99.2% Fe alloy is shown in (c). It consists of only one microconstituent, the eutectoid pearlite. The microstructure at 70°F is the same as at 1332°F, which is 1°F below the eutectoid temperature.

The microstructure at 1332°F and 70°F of a slowly cooled 0.4% C-99%.6% Fe alloy is shown in (d). Since this composition lies in the two-microconstituent region of $\alpha + P$, the microstructure must contain both α and P. The inverse lever rule is employed to calculate the amount of each microconstituent present, with the following results, where the tie line is again ab:

$$P = \frac{pa}{ba} = \frac{0.4 - 0}{0.8 - 0} = 1/2, \text{ or } 50\%$$

$$\alpha = \frac{bp}{ba} = \frac{0.8 - 0.4}{0.8 - 0} = 1/2, \text{ or } 50\%$$

The composition of point *a* is taken as 0% C because the error introduced with this simplification for plain carbon steels is negligible since the impurities that are present in steel lower the solubility of carbon in α iron. The microstructure for this same alloy at a temperature of 1334°F would consist of the same 50% α, but the pearlite would now be austenite of 0.8% C.

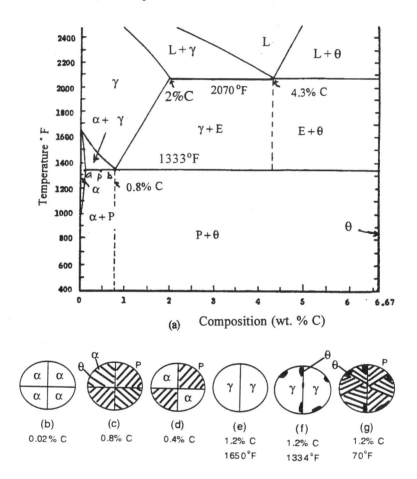

Figure 4-4. The Fe-C microconstituent diagram and some typical microstructures.

Microconstituent Diagrams and Microstructures

Consider next an alloy of 1.2%C - 98.8%Fe. At 1650°F it is in the one-microconstituent region, labeled γ so the microstructure consists of 100% austenite having 1.2% C dissolved interstitially. The microstructure is sketched in (e), with only two grains to indicate a large grain size at this temperature. After slowly cooling to 1334°F, the microstructure appears as in (f), indicating that some cementite has precipitated in the austenite grain boundaries. The amount of θ present is

$$\frac{1.2-0.8}{6.67-0.8} = \frac{0.4}{5.87} \text{ or } 7\%$$

Since the amount of P present is so small, it does not form large grains but occurs as small particles concentrated in the grain boundaries as shown.

After slowly cooling past the lower critical temperature, 1333°F, where the transformation of austenite to pearlite occurs, the microstructure appears as in (g). The significant feature illustrated here is that each grain of austenite transforms into several grains of pearlite. At 1332°F, and also at 70°F, the θ remains unchanged from its condition at 1334°F.

RULES OF MECHANICAL PROPERTIES

A phase diagram contains no information regarding the mechanical properties of the alloys formed from its components. Even the concept that the tensile strength of a material is related to its melting temperature is not always valid because there are many cases in which a comparison of two materials shows that the one with the lower melting temperature has a higher room-temperature tensile strength. For example, nickel has a melting temperature of 2651°F and a tensile strength of 46 ksi, whereas platinum has a melting temperature of 3224°F and a tensile strength of only 18 ksi; and both have FCC lattice structures. Even if the mechanical properties of the two phases α and β of a binary phase diagram are known, it is not possible to determine by any means, other than experimental tests, what the mechanical properties are of all the alloys that contain various combinations of the two phases. To determine the mechanical properties, a separate set of tests has to be performed for each composition.

The justification for microconstituent diagrams — the reason for their importance to the practicing engineer working with mechanical systems — is that they make it possible to calculate, with reasonable reliability, the mechanical properties of any alloy from the known properties of the individual

102   Chapter 4

microconstituents. For example, one cannot know the strength of a 40% Sn-60% Pb alloy on the basis of the strength of the α and β phases of the phase solid regions shown in Figure 3-7. But a reasonable prediction can be made for the strength of this alloy on the basis of the strength of the α and E eutectic microconstituents, as explained below. Likewise, the strength of annealed AISI 1040 steel on the basis of the strength of the ferrite and cementite phases cannot be calculated, but it can be calculated from the strengths of the ferrite and pearlite microconstituents with at least as much reliability as having the specimen tested in a commercial testing laboratory.

The rules for determining mechanical properties on the basis of microstructures are as follows.   For ferrous materials, rules 3–8 also can be applied.

1. The mechanical properties of any alloy are a weighted average of the properties of the individual microconstituents. That is:
$(MP)alloy = (f)MC_1 \times (MP)MC_1 + (f)MC_2 \times (MP)MC_2$ where (MP) is any mechanical property and (f) is the fraction of any microconstituent (MC) present.
2. The maximum amount of any microconstituent that can be present is the equilibrium amount.
3. The carbon content of pearlite is always the same as the carbon content of the eutectoid.
4. The carbon content of ferrite is assumed to be zero.
5. Ferrite (for hypoeutectoid steel) or cementite (for hypereutectoid steels) transforms from the austenite first, that is, before any pearlite or martensite forms.
6. If any pearlite is present, then the amount of ferrite present is the same as the equilibrium amount.
7. If only ferrite and martensite are present, then the amount of ferrite can vary between 0% and the equilibrium amount.
8. The carbon content of martensite is such that the product of the fraction martensite present times its carbon content equals the carbon content of the steel (when the other microconstituent is α).

Consider again the mechanical properties of annealed AISI 1040 steel. The average or typical properties of ferrite (see Table 6-1, p. 155) are: 80 Brinell hardness, 40 ksi tensile strength, 75% reduction of area, and 30×10^3 ksi modules of elasticity. The typical properties for coarse lamellar pearlite are: 240

Brinell hardness, 120 ksi tensile strength, 30% reduction of area, and 30×10^3 ksi modules of elasticity. Since the microstructure of annealed 1040 steel consists of 50% ferrite and 50% coarse pearlite, its mechanical properties are calculated from the above rule as follows:

$$\begin{aligned}
\text{Brinell hardness} &= 1/2 \times 80 + 1/2 \times 240 = 160 \,\text{kg}/\text{mm}^2 \\
\text{Tensile strength} &= 1/2 \times 40 + 1/2 \times 120 = 80 \,\text{ksi} \\
\text{Reduction of area} &= 1/2 \times 75 + 1/2 \times 30 = 52\% \\
\text{Modulus of elasticity} &= 1/2 \times 30 + 1/2 \times 30 = 30 \times 10^3 \,\text{ksi}
\end{aligned}$$

In similar fashion the properties for an annealed 1020 (0.20% C) steel, whose microstructure consists of 75% ferrite and 25% pearlite, are calculated as follows:

$$\begin{aligned}
\text{Brinell hardness} &= 3/4 \times 80 + 1/4 \times 240 = 120 \,\text{kg}/\text{mm}^2 \\
\text{Tensile strength} &= 3/4 \times 40 + 1/4 \times 120 = 60 \,\text{ksi} \\
\text{Reduction of area} &= 3/4 \times 75 + 1/4 \times 30 = 63\% \\
\text{Modulus of elasticity} &= 3/4 \times 30 + 1/4 \times 30 = 30 \times 10^3 \,\text{ksi}
\end{aligned}$$

The mechanical properties of any annealed steel can be determined just as easily. Chapter 5 presents the techniques and data necessary to predict the microstructures after non-equilibrium cooling, which make it possible to calculate the strength of *any* steel after *any* heat treatment (and with as much reliability as looking up the values in a handbook).

FERROUS EQUILIBRIUM MICROCONSTITUENTS

It is appropriate at this time to define the ferrous microconstituents that are present under equilibrium conditions and to explain in detail the pearlite reaction. Although these microconstituents were mentioned in the previous section, they were not explained or defined. Their mechanical properties are tabulated in the Chapter 6, along with the non-equilibrium microconstituents.

Austenite

Austenite is a solid solution of carbon dissolved interstitially in a face-centered cubic lattice structure of iron. It exists in the ordinary steels only at temperatures above the lower critical temperature. However, on rapid cooling, austenite may exist for a short period of time (a few seconds to a few minutes) at temperatures down to approximately 600°F, as explained in Chapter 5. In some of the austenitic stainless steels it exists, and is stable, at cryogenic temperatures. It is also referred to as gamma iron, or γ iron.

Ferrite

Ferrite is a solid solution of carbon dissolved interstitially in the lower temperature body-centered cubic lattice structure of iron. The amount of carbon that is soluble in ferrite is very small, 0.008% at room temperature and 0.025% at the eutectoid temperature. This decrease in the solubility of carbon at the lower temperature results in a precipitation hardening effect that is significant for the very low carbon steels, less than 0.1% C, but is insignificant for the other steels in comparison to the strengthening effect of obtaining a supersaturated solid solution (called primary martensite and discussed in the next chapter) by heat treating. Ferrite is also referred to as alpha iron, or α iron.

Cementite

Cementite is an intermetallic compound of iron and carbon having the chemical formula Fe_3C. It has an orthorhombic lattice structure with covalent bonds. This type of bonding makes it very hard and brittle — it has no ductility and by itself cannot be plastically deformed. It is the microconstituent that imparts wear-resistance to the hypereutectoid steels. It is also referred to as iron carbide, or just carbide.

Pearlite

Pearlite is a mechanical mixture of ferrite and cementite, usually in lamellar form, and of eutectoid composition. The term "mechanical mixture" means that the two phases are physically distinct and can actually be mechanically separated, as opposed to a chemical mixture, where the phases lose their individual identities. The expression "of eutectoid composition" is relevant because it is not necessary for a steel to have a 0.8% carbon content in order to have a microstructure that consists of only the one microconstituent, pearlite. For example, both a 0.8% C plain carbon steel and a 0.4% C alloy steel that contains 1.25% Mo are 100% pearlite (see Figure 4-5). That is, their microstructure is entirely pearlite.

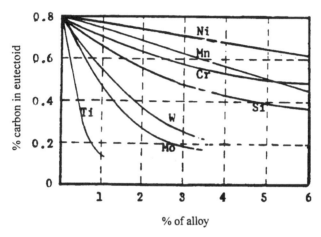

Figure 4-5. Effects of alloy content on the carbon content of the eutectoid pearlite.

When alloys are added to steel, they dissolve in the austenite (they replace iron atoms in the austenite lattice structure) at temperatures above the critical temperature, and as a consequence they lower the amount of carbon that can dissolve interstitially in the austenite. Figure 4-5 shows how much the common alloying elements lower the solubility of carbon in austenite at the eutectoid temperature. For example, when 1.25% of molybdenum is added to steel, the carbon content of the eutectoid is lowered to 0.40%; the addition of 1% of chromium lowers the eutectoid composition to 0.70% C. When small amounts of several alloys are added to a steel, they have an additive effect of lowering the carbon content of the eutectoid. Thus 0.5% Mo lowers the carbon content by 0.18% (from 0.80% to 0.62%) and 1% Cr lowers it by 0.10% (from 0.80% to 0.70%), so a steel having both these alloys would have a eutectoid composition that would be lower than the 0.80% by 0.28%; that is, its eutectoid composition would be 0.52% C. Any carbon content greater than this would make the alloy steel hypereutectoid.

The curves in Figure 4-5 can be approximated by the following relations, where ΔC is reduction in the carbon content of the eutectoid.

$$\Delta C_{Ni} = -0.030 \times \%Ni \left(\text{up to } 6\% \right)$$
$$\Delta C_{Mn} = -0.061 \times \%Mn \left(\text{up to } 6\% \right)$$
$$\Delta C_{Cr} = -0.080 \times \%Cr \left(\text{up to } 3\% \right)$$
$$\Delta C_{Si} = -0.125 \times \%Si \left(\text{up to } 3\% \right)$$

$$\Delta C_W = -0.222 \times \%W \left(\text{up to } 1\% \right)$$
$$\Delta C_{Mo} = -0.33 \times \%Mo \left(\text{up to } 1\% \right)$$
$$\Delta C_{Ti} = -1.00 \times \%Ti \left(\text{up to } 0.5\% \right)$$

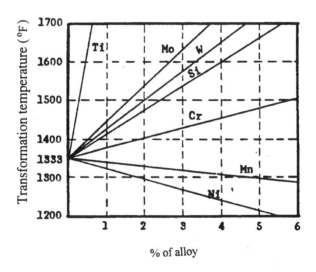

Figure 4-6. Effect of alloy content on the austenite-to-pearlite transformation temperature (lower critical temperature).

Figure 4-6 illustrates the effect that the common alloying elements have on the lower critical temperature. The alloys shown, except for nickel and manganese, raise the eutectoid temperature. Thus it is apparent that a steel containing 2% Si and no other alloys must be heated to 1475°F to become

Microconstituent Diagrams and Microstructures 107

austenitic. The effects of small amounts of several alloys are additive. Thus 1% Ni cancels out the effect of 0.25% Mo on changing the critical temperature.

The pearlite reaction. The *pearlite reaction* refers to the mechanisms by which austenite transforms to pearlite. An understanding of the pearlite reaction makes it easier to understand the concepts of hardenability that are presented in the following Chapter 5. The discussion presented here is valid for any eutectoid; it is also valid for all eutectic reactions given that the two final solid phases nucleate and grow from a liquid rather than a third solid phase.

Austenite transforms to pearlite by a diffusion, nucleation, and growth process that is similar to that which occurs when a liquid transforms to a solid. As such, it is time dependent as well as temperature dependent. Figure 4-7 illustrates how this transformation takes place. Figure 4-7(a) shows the microstructure of a eutectoid steel at the first instant it reaches 1333°F after cooling from a higher temperature. Some nuclei of ferrite and cementite form at the austenite grain boundaries and are shown as the small black specks. In this context a nucleus is a small cluster of atoms that have arranged themselves into a crystalline lattice structure. Figure 4-7(b) is an enlargement of a grain boundary between two austinite grains, γ_1 and γ_2, showing two nucleation sites. Figure 4-7(c) is a further enlargement of the lower nucleation site.

The first step in the transformation process is the diffusion of the carbon in the austenite, which has a uniform composition of 0.8% C, to regions of high carbon content and adjacent regions of low carbon content. When the carbon content decreases to 0.025% at a small part of the grain boundary, the lattice structure changes from FCC to BCC. Immediately adjacent to this low-carbon BCC nucleus, the carbon content increases to 6.67% and a nucleus of cementite forms. More nuclei form at the original grain boundary adjacent to the first ones and the direction of nucleation is parallel to the grain boundary. Each nucleus grows outward, perpendicular to the grain boundary, as shown in Figure 4-7(c), and the carbon continues to diffuse toward the cementite nuclei allowing the growth to continue. Soon two nuclei growing from the same grain boundary meet, resulting in a grain boundary between two pearlite grains that has grown out of the same austenite grain. Eventually all of the austenite transforms to pearlite, as shown in Figure 4-7(e).

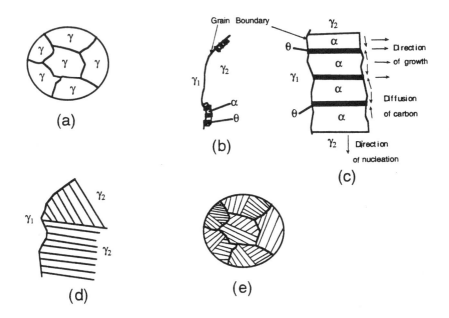

Figure 4-7. The pearlite (eutectoid) reaction. (a) Austenite upon first reaching 1333°F. (b) An enlargement of a nucleation site, showing a grain boundary between austenite grains. (c) A further enlargement of (b), showing direction of diffusion, nucleation, and growth. (d) Same location as (b) only a short time later. (e) Microstructure at 1333°F, after complete transformation.

The number and size of grains that form when a liquid transforms to a solid, or when one solid solution transforms to one or two other solid solutions, depends upon the cooling rate, or, to be more exact, the degree of undercooling, which is the difference between the equilibrium transformation temperature and the actual temperature of the material during transformation. When a liquid is cooled below its melting temperature or a solid is cooled below its equilibrium transformation temperature, many nuclei begin to form at various sites at the same time. Figure 4-8 illustrates how the nucleation rate is affected by the degree of undercooling. For a small amount of undercooling, just to the right of the equilibrium temperature T_E in Figure 4-8, only a few nuclei form and they grow to a large size, so the growth rate G is large. But with more rapid cooling and a greater degree of undercooling, such as at T_2 in Figure 4-8, the rate of nucleation is high and the rate of growth of the grains is low. In some cases, if the rate of cooling is too great, there is insufficient time for diffusion to occur, and as a result the amount of nucleation decreases and may even be zero.

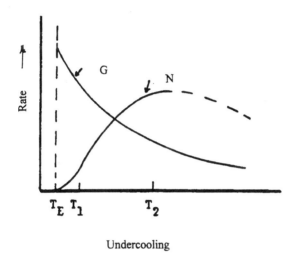

Figure 4-8. Rate of nucleation, N, and rate of growth, G, as a function of the degree of undercooling. T_E is the equilibrium temperature.

Specific volume changes. As austenite transforms to the lower temperature equilibrium microconstituents, the specific volume increases. Likewise, on heating, there is a similar decrease in specific volume, due to more efficient packing of atoms. This is illustrated in Figure 4-9. Curve (a) shows this change in specific volume occurring at 1670°F for the $\gamma \rightarrow \alpha$ transformation. Curve (b) shows this change at the constant temperature of 1333°F for the $\gamma \rightarrow P$ transformation temperature. For example, a 1040 steel (0.4% C) as indicated by curve (c) has a fairly constant specific volume in the temperature range of 1500°F to 1333°F, where the partial transformation of austenite to ferrite is offset by the thermal contraction associated with this drop in temperature. Then at 1333°F the remaining 50% of austenite transforms at constant temperature to pearlite.

In Figure 4-9, the curve for the 0.80 C steel is drawn above that of the 0% C steel because the unit cell side length, or the lattice parameter, of austenite increases as the amount of carbon dissolved interstitially in it increases. The lattice parameter a in angstrom units is equal to $3.555 + 0.44 \times \% C$.

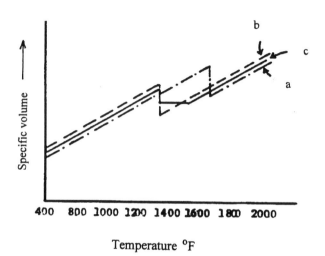

Figure 4-9. Specific volume changes of ferrous microconstituents: (a) ferrite, 0% C; (b) pearlite, 0.8% C; (c) 1040 steel, 0.4% C.

Although not shown in Figure 4-9, if the 0.8% C and 0.4% C steels were quenched to form martensite, the $\gamma \rightarrow M$ transformation would occur over the temperature range of 600°F to 200°F, with the resulting expansion occurring at the lower end of the temperature range. It is this change in specific volume that causes steel to crack during the rapid heating or cooling associated with the heat treating and welding processes.

REFERENCES

1. Brick, R.M., R.B. Gordon and A. Phillips, *Structure and Properties of Alloys,* McGraw-Hill, New York, 1965.
2. Samans, C-H. *Metallic Materials in Engineering,* Macmillan, New York, 1963.
3. Murray, G.T. *Introduction to Engineering Materials,* Marcel Dekker, New York, 1993.
4. van Vlack, L.H. *Elements of Materials Science,* Addison-Wesley, MA, 1989.
5. Flinn, R.A. and P.H. Trojan, *Engineering Materials and Their Applications,* Houghton Mifflin, Boston, 1990.
6. Shackelford, J.F. *Material Science for Engineers,* Macmillan, New York, 1992.
7. ASM, *Relation of Properties to Microstructure,* American Society of Metals, Ohio, 1954.
8. Datsko, J. *Material Properties and Manufacturing Processes,* John Wiley & Sons, New York, 1967.

Chapter 5

NON-EQUILIBRIUM DIAGRAMS AND MICROCONSTITUENTS

INTRODUCTION

Although a knowledge of equilibrium microstructures is essential to a thorough understanding of the control of the mechanical properties of all materials, equilibrium microstructures are found only infrequently in real structural components. Since much higher strengths can be achieved by the various non-equilibrium heat treating processes, nearly all structural parts have non-equilibrium microstructures. The non-equilibrium microstructures for many materials can be predicted solely on the basis of their equilibrium phase diagrams plus some simple experimental cooling rate and aging temperature data. This is true of the aluminum alloys as well as the exotic age-hardenable alloys. The situation is much more complicated for those alloy systems that undergo allotropic transformations or that have eutectoid reactions, as described in the previous chapter. Among these more complicated alloy systems are titanium alloys, aluminum bronze, and, of course, steel.

This chapter shows how to interpret and use non-equilibrium diagrams and microconstituents to control the mechanical properties of materials. Since steel is the most important structural material and much more data is available for this family of materials than any other, it will be used as an example to demonstrate the types of non-equilibrium diagrams and microconstituents that are available to the engineer who wants to specify the optimal material and treatment. The same techniques that are used in this chapter to select and specify steel are applicable to any other materials system, though, unfortunately for many of these systems the required diagrams are not available.

DIFFUSION AND TRANSFORMATION RATES

All equilibrium phase transformations occur by the diffusion, nucleation, and growth process that was described in Chapter 4. Whenever a non-equilibrium phase or microconstituent occurs, it is because the cooling rate has been so fast that there was insufficient time for diffusion and nucleation to take place. It is necessary to have an appreciation of the diffusion processes in order to understand when and why non-equilibrium phases and microconstituents occur.

Diffusion is a movement of atoms, or molecules, that is caused by a combination of thermal energy and a concentration gradient that tends to make the material homogeneous in composition. Slip is not a diffusion process, nor is the martensite reaction, because they involve the mass movement of a large number of atoms resulting from an applied mechanical stress rather than a concentration gradient. Diffusion, when a material is in a gaseous or liquid state, occurs rapidly because of the rapid and random movement of the molecules of a fluid. But diffusion occurs very slowly in solid materials, particularly at temperatures well below the melting point, because the atoms must move from one lattice site to another or from one interstice to another. As explained in the first chapter, the atoms of a solid material at a given temperature are in their lowest energy level when they occupy sites at the lattice points of a crystal. In order to move an atom from one lattice point to another, the energy level of the atom must be raised while it is in transit, as illustrated in Figure 5-1. This energy is called the *activation energy* and it comes from the thermal excitation or increased amplitude of vibration of the atom.

It is believed that diffusion cannot occur in a solid having a perfect crystal lattice. It is the presence of point defects, principally vacancies and interstitials, that permits the atoms to move rather freely throughout the lattice structure. Most diffusion occurs by the vacancy mechanism illustrated in Figure 5-2(a). This is particularly true for *self-diffusion*, which is the diffusion of like atoms in a solid of a pure element. This type of diffusion has been experimentally verified by making the surface layer of atoms of a solid such as copper radioactive, and then tracing the position of the radioactive atoms in the solid as a function of time. All three mechanisms illustrated in Figure 5-2 are active for *interdiffusion*, which is the diffusion of unlike atoms in a crystalline solid. Both the vacancy and the interstitial mechanisms are responsible for the diffusion of carbon in γ iron. The atom interchange mechanism shown in Figure 5-2(c) occurs very infrequently.

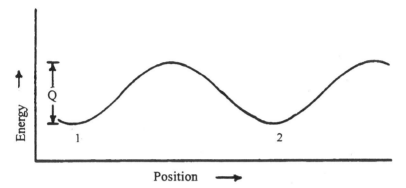

Figure 5-1. Energy level of an atom as it moves from one lattice site to an adjoining site. Positions 1 and 2 are neighboring equilibrium lattice sites. Q is the activating energy necessary for an atom to transfer from one site to a neighboring site.

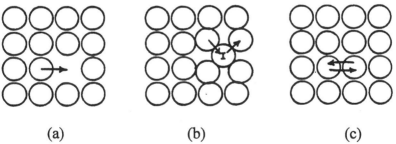

Figure 5-2. Mechanisms of diffusion in crystalline solids: (a) movement of vacancies, (b) movement of interstitial atoms, (c) atom interchange.

Diffusion that occurs throughout the volume of a crystal is called *volume diffusion*. As the name implies, *grain boundary diffusion* refers to the diffusion that occurs along the grain boundaries of a polycrystalline solid. This kind of diffusion occurs because the energy level of the atoms in the vicinity of the grain boundaries is higher than that of the central regions of the grains and there are more lattice defects at the grain boundaries than at the bulk of the material. Diffusion that occurs over a surface, such as the two surfaces of a crack, is called *surface diffusion*.

Diffusion is due to a concentration gradient and tends to make a material homogeneous. Figure 5-3 illustrates the atom movements and concentration gradients that exist during diffusion. The atoms marked X could be radioactive atoms of copper on the surface of a copper crystal with (I) being the original interface between the radioactive and normal atoms, or they could be zinc or nickel atoms plated onto a copper crystal, or (I) could be the interface between a smooth flat face of a nickel crystal and a similar one of a copper crystal.

There are three quantitative relationships that define the number of atoms that diffuse through a solid because of a concentration gradient. The first two are known as Fick's laws. The first law refers to one-dimensional steady-state and diffusion is usually expressed as

$$\frac{dm}{dt} = -AD\frac{dC}{dx} \qquad (5\text{-}1)$$

where dm is the number of atoms that cross a plane that is perpendicular to the direction of diffusion in the time interval dt (sec), A is the area (cm^2) of the plane through which the atoms are diffusing, D is the *diffusion coefficient* or *diffusivity* (cm^2/sec), dC is the difference in concentration (atoms/cm^3) of the diffusing atoms at two points separated by the distance dx (cm) in the direction of diffusion. The negative sign indicates that the atoms diffuse in a direction opposite to that of the concentration gradient dC/dx.

Fick's second law applies to the more common unsteady-state conditions in which the concentration gradient dC/dx varies with time, such as the carburization of steel. The simplified form of Fick's second law, which assumes that the diffusivity D is constant at a given temperature and does not vary with composition, is

$$\frac{dC}{dt} = D\frac{d^2C}{dx^2} \qquad (5\text{-}2)$$

Non-Equilibrium Diagrams and Microconstituents 117

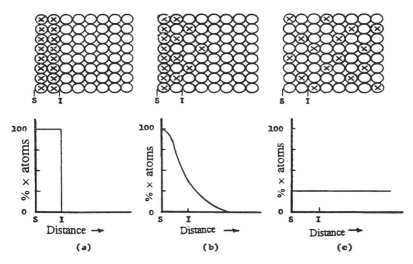

Figure 5-3. Diffusion and concentration gradients for self-diffusion and interdiffusion. Atoms marked X may be the same as the unmarked atoms, except for being radioactive, or they may be a different element such as zinc, with the remainder being copper. S is a surface and I is the original interface. (a) At zero time. (b) At an intermediate time. (c) After homogenization, with the time depending upon the temperature.

The diffusivity, D, is a measure of the rapidity with which atoms move through a crystalline solid, and it is very strongly affected by the temperature of the solid since the atom's ability to change lattice sites is directly related to its thermal vibrations. The diffusivity of metals is related to temperature by the Arrhenius rate equation

$$D = D_o\, e^{-Q/RT} \tag{5-3}$$

where Do (cm^2/sec) is a constant that is not related to the diffusing atom's amplitude of vibration (i.e., not affected by temperature) but is probably related to the atom's frequency of vibration, e is the base of the natural logarithm, Q is the activation energy (cal/mole), R is the gas constant and has the value 1.987

118   Chapter 5

cal/mole K, and T is the temperature in degrees Kelvin. This equation is generally displayed graphically by plotting log D vs. log (1/T).

Sometimes it is more convenient to use the rate equation in logarithm form, which is obtained by taking logarithms of both sides of the above exponential form as follows:

$$\ln D = \ln Do - (Q/RT) \ln e$$

or (5-4)

$$\ln D = \ln Do - Q/RT$$

since $\ln e = 1$. (ln is the symbol for logarithm to the base e, $\log_e$; also, $\ln x = 2.3 \log_{10} x$).

Since the temperature is in the exponent of the rate equation, it has a very pronounced effect on the diffusivity, as can be seen from the following examples.

The proportionality constant Do for carbon in γ iron is 0.21 cm²/sec and the activation energy Q is 34,000 cal/mole. Thus the diffusivity D at 1333°F (995°K) is calculated as follows:

$$D = \frac{0.21}{34,000/e^{1.987T}} = \frac{0.21}{e^{17.18}} = 7.53 \times 10^{-9} \, cm^2/sec$$

The value of 1333°F was selected since it is the lowest temperature at which γ iron is stable. By doubling the temperature to 2666°F (1733K), the diffusivity is increased to $1.1 \times 10^{-5} cm^2/sec$, which is 10,000 times faster. This clearly shows the great effect temperature has on diffusion.

Another interesting comparison is how much the temperature has to be raised above 1333°F in order to double the diffusion coefficient, that is, to raise it from $7.53 \times 10^{-9} cm^2/sec$ to $15.06 \times 10^{-9} cm^2/sec$. This is determined as follows:

$$D = \frac{0.21}{e^x} \quad \text{where} \quad x = \frac{Q}{RT}$$

Non-Equilibrium Diagrams and Microconstituents

or

$$e^x = \frac{0.21}{D} = \frac{0.21}{15.06 \times 10^{-9}} = 1.39 \times 10^7$$

so

$$x = 16.3$$

and

$$T = \frac{Q}{Rx} = \frac{34,000}{1.987 \times 16.3} = 1050K \text{ or } 1430°F$$

Thus the diffusivity of carbon in γ iron can be doubled by raising the temperature from 1333°F to 1430°F.

The diffusion of carbon is faster in α iron since the BCC structure is a less efficient packing of atoms than the FCC. This can be seen from the fact that the constant Do is 0.008 cm^2/sec and the activation energy Q is 18,000 cal/mole for this allotropic form of iron. Therefore at 1333°F the diffusivity is 8.7×10^{-7} cm^2/sec, which is 100 times faster than in γ iron at this temperature. It is unfortunate for the carburization processes that α iron does not exist at the higher carburizing temperatures.

The transformation times or reaction rates for most solid-solid transformations such as the precipitation or recrystallization processes follow the same Arrhenius rate equation. In these cases it is written as

$$r = Ae^{-Q/RT} \quad \text{or} \quad \frac{1}{t} = Ae^{-Q/RT} \tag{5-5}$$

where r is the reaction rate (reciprocal of time), t is the time in seconds, and A is a proportionality constant. Q, R, and T are the same as previously defined. To determine the numerical value for the constant A and the activation energy Q, it is necessary simply to experimentally measure the time for the reaction to occur (or begin) at two temperatures. After A and Q are known, the reaction time at any other temperature can be easily calculated.

The Arrhenius rate equation is not valid at temperatures just below the equilibrium transformation temperatures. The driving force for precipitation to occur is the difference in free energy between the supercooled metastable phase and the precipitated phase. But additional energy, activation energy, is necessary to form the new surfaces created when new particles nucleate from a

120 Chapter 5

liquid or solid solution. The difference in free energy depends upon the degree of supercooling below the equilibrium transformation temperature. The driving force is very low just below the equilibrium transformation temperature, so few nuclei form and precipitation is slow. The driving force increases at somewhat lower temperatures and precipitation occurs more rapidly. However, the diffusion rate decreases at lower temperatures and so the two effects counterbalance each other. Figure 5-4 illustrates how the reaction time for precipitation varies with temperature. The portions of the curves below point x follow the rate equation fairly well, but the portions near the equilibrium transformation temperature do not. These curves are called "c" curves because of their shape. That portion of the curve where the reaction is shortest is called the *knee* or the *nose* of the curve.

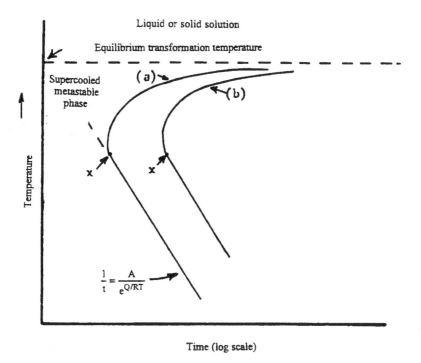

Figure 5-4. Reaction time curve for precipitation. (a) Start of transformation. (b) End of transformation.

NON-EQUILIBRIUM PHASES AND MICROCONSTITUENTS

It is possible to obtain non-equilibrium phases at room temperature under some conditions. The most common instances occur with materials that exist under equilibrium conditions as two phases at room temperature but exist as only a single phase at a high temperature. When such a material is rapidly cooled from the high temperature, there may be insufficient time for diffusion and nucleation to occur, and so a non-equilibrium phase, or microconstituent, may result. For example, if a 5% Cu-95% Al alloy (see Figure 3-9) is quenched from 900°F, a supersaturated solid solution of aluminum containing 5% Cu will exist at room temperature for many hours, or even days, whereas the equilibrium phases are a solid solution containing only about 0.5% Cu plus the compound $CuAl_2$. Or, if a 16% aluminum bronze is rapidly quenched from 1100°F, where it is a β solid solution, a martensitic type of solid solution, rather than the two equilibrium phases, is present at room temperature.

Medium and Fine Pearlite

The most important non-equilibrium microconstituents are the ones that occur with the ferrous alloys. When a plain carbon steel of eutectoid composition is very slowly cooled (annealed), the equilibrium microconstituent that results is coarse lamellar pearlite, or sometimes spheroidized pearlite. However, if a small piece of this same steel is air cooled, more nuclei of ferrite and cementite form, but with less growth and the resulting microconstituent will be platelets of medium thickness, called *medium pearlite*.

If the cooling rate is faster, such as with an oil quench, the platelets of ferrite and cementite will be the very fine microconstituent called *fine pearlite*.

Martensite

Martensite is a supersaturated solid solution of carbon dissolved interstitially in either a body-centered tetragonal or a body-centered cubic structure of iron. The original martensite that forms is a light- or white-colored microstructure that has a tetragonal structure and is called *primary* martensite. In comparison to the cubic structure of α iron, the tetragonal structure has one side length that is longer and two that are shorter. The ratio of side lengths (longest side to

122 Chapter 5

shortest side) of the tetragonal martensite structure varies from 1.08 when the carbon content is 1.8% to 1.00 when the carbon content is zero.

When primary martensite is heated to about 225°F, the body-centered tetragonal lattice transforms to a body-centered cubic structure by having one axis shortened and the other two lengthened to make all three equal. This change in structure is accompanied by a very slight decrease in volume, less than 0.1%, and a relief of residual quenching stress. This form of martensite is called *secondary* martensite or yellow martensite since its microstructure has a yellowish tint.

Both primary and secondary martensite have a microstructure that, in appearance, is acicular or needle-like rather than having equiaxial grains. The needles appear to intersect each other at 60° and 120° angles (see Figure 5-5) which correspond to the intersections of the (111) planes in the austenite grains, as discussed in the following section.

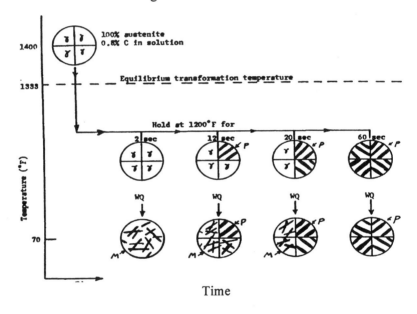

Figure 5-5. Isothermal transformation of eutectoid steel (0.8% C) at 1200°F. γ = austenite, P = pearlite, M = martensite, WQ = water quench.

Martensite that is reheated to temperatures above 250°F but less than the lower critical temperature (approximately 1333°F) is called *tempered* martensite. Reheating to temperatures within this range causes some carbon to diffuse and precipitate out of the solid solution as either a complex iron carbide (not Fe_3C) at the lower tempering temperature or as cementite at the higher tempering temperatures. Thus tempered martensite consists of two phases or microconstituents, iron plus carbides. On the other hand, both primary and secondary martensite consist of only one phase or microconstituent.

The martensite reaction. Martensite does not form by the diffusion, nucleation, and growth process that is time dependent, but instead by a diffusionless "shear" mechanism that is only temperature dependent. Thus if the above-mentioned steel were cooled at a rate faster than that which produces fine pearlite (water quenching small sections of plain carbon steel is sufficiently fast), there would be insufficient time for the diffusion of carbon and nucleation of ferrite and cementite to occur. When the temperature of the supercooled austenite or *retained* austenite (the term used to designate austenite that exists below the lower critical temperature) drops to about 430°F, a small amount of austenite transforms instantaneously to primary martensite. This temperature at which austenite first begins to transform to martensite is called the martensite start temperature and is labeled M_s on the isothermal transformation diagrams such as Figure 5-6. If the steel were held at this temperature for several hours, no additional austenite would transform to martensite. Upon cooling to a temperature of 330° about one-half of the austenite would transform to martensite. Even after several hours at 330°F, the steel would still consist of 50% austenite and 50% martensite. Further transformation occurs only by decreasing the temperature. By the time the temperature drops to 220°F, the M_f or martensite finish temperature, all of the retained austenite has transformed to primary martensite so that at room temperature only one phase, primary martensite, exists.

X-ray studies made of martensite and the austenite from which it has been transformed show that the [111] direction in the martensite is parallel to the [110] direction of the austenite and that the (110) planes in the martensite are parallel to the (111) planes of the austenite. On the basis of this evidence it is believed that the transformation from austenite to martensite takes place by a shearing process similar to twinning. Successive (111) planes of the austenite grain become displaced relative to one another by a small amount, similar to the twinning process. The transformation mechanisms for the high carbon steels and the low carbon steels appear to be different.

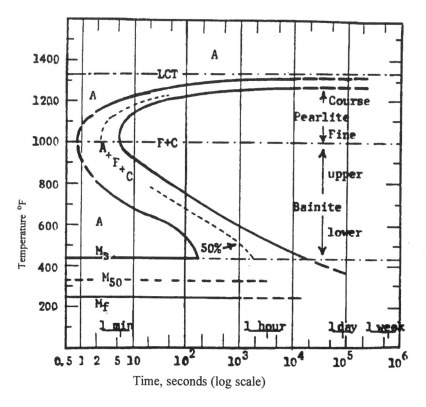

Figure 5-6. Isothermal transformation diagram for a typical AISI 1080 steel:
A = austenite, B = bainite, C = carbide, F = ferrite, M = martensite.

Bainite

Bainite is the microconstituent or microstructure that results when a steel is isothermally transformed from austenite at temperatures between the minimum pearlite formation temperature and the start of the martensite transformation temperature. In appearance the microstructure of bainite is feathery, acicular, or needle-like and resembles martensite in this respect. On the other hand, bainite forms by a diffusion, nucleation, and growth process similar to the pearlite

Non-Equilibrium Diagrams and Microconstituents

reaction and not by the "shear" mechanism of the martensite reactions. The differences between the pearlite reaction and the bainite reactions are simply which precipitates first, the ferrite or the cementite, and where. It is believed that in the pearlite reaction cementite precipitates first, followed by ferrite at the austenite grain boundaries, with growth perpendicular to the grain boundaries. In the bainite reaction it is believed that plates of supersaturated ferrite form along the (111) planes of the austenite grains and that particles rather than platelets of carbide begin to precipitate out of the supersaturated ferrite, with the rate of precipitation depending upon the temperature. Bainite is not normally found in plain carbon steels because the pearlite nose extends to very short reaction times, as shown in Figure 5-6.

The mechanical properties of bainite vary between those of fine pearlite and primary martensite, depending upon the temperature at which it is formed. The mechanical properties of all of the ferrous microconstituents are listed in Table 6-1 (p.156).

ISOTHERMAL TRANSFORMATION DIAGRAMS

Isothermal transformation diagrams are curves plotted on time vs. temperature coordinates that indicate when a supercooled solid solution will transform either to two equilibrium phases or else to a fourth metastable phase. The only important isothermal transformation diagrams in use today are those for steel, and the best source for them is the *Atlas of Isothermal Transformation Diagrams,* published by the United States Steel Corporation. Since these diagrams are very helpful in understanding the effect of time and temperature on the transformation of austenite, as well as in the selection of the optimal heat treatment for a particular steel, they are discussed in some detail in this text. The isothermal transformation diagrams are also called *TTT curves* (for they represent time, temperature, and transformation), *C curves* because of the shape of the pearlite transformation part of the diagram, and *S curves* because of the shape of the entire curve including the martensite start and finish lines. The curves are obtained from experimental isothermal heat treatments as described below.

Figure 5-5 serves as a convenient means to illustrate how the TTT curves are obtained. In this case an AISI 1080 steel is being studied. Four discs about 1" diameter and 1/16" thick are placed in a furnace at a temperature above the critical temperature, say, 1400°F. After they have become austenitized (100% homogenized austenite), which requires a minimum of 30 minutes or 1

126 Chapter 5

hr/in. thickness, one piece is rapidly taken out of the furnace and plunged beneath the surface of a lead bath that is maintained at a temperature of 1200°F. Two seconds after being immersed in the lead the piece is then withdrawn and water quenched (cooled to room temperature almost instantly). Following this treatment the piece can be subjected to a hardness, microstructural, or X-ray study. Such a study would indicate that the room temperature hardness is 67 Rc (Rockwell C) or nearly 800 H_B (Brinnel hardness) and that the microstructure is 100% martensite. This means that the steel was austenite after 2 seconds at 1200°F because once austenite transforms to pearlite there is no other change upon further cooling.

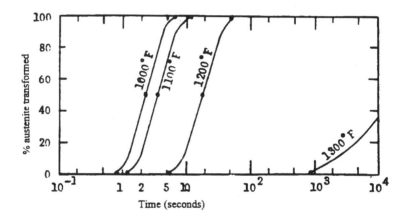

Figure 5-7. Austenite transformation rate at constant temperature.

In a similar manner, a second piece is rapidly plunged into the lead and held there for 12 seconds whereupon it is withdrawn and water quenched. The final room temperature hardness is about 62 Rc (695 H_B) and the microstructure is 25% fine pearlite and 75% martensite.

A third piece is water quenched (cooled to room temperature almost instantly) after 20 seconds at 1200°F. Its room temperature hardness if 57 Rc (590 H_B) and its microstructure consists of 50% fine pearlite and 50% martensite. The last piece is held in the lead bath for 60 seconds before water quenching. Its final hardness is 41 Rc (380 H_B) and its microstructure is all fine pearlite.

If one were to study additional specimens that were held at 1200°F for times between 2 and 12 seconds, and also between 20 and 60 seconds, it would be possible to determine at what time the austenite began to transform to $\alpha + \theta$ and at what time the transformation was complete. These data can then be plotted as a curve of percent austenite versus time as shown in Figure 5-7 for the 1200°F transformation temperature. By repeating this study at several temperatures between 1333°F (the maximum temperature at which pearlite exists) and 1000°F (the minimum temperature at which austenite transforms to pearlite in an AISI 1080 steel), a family of curves such as those in Figure 5-7 can be drawn.

The above method cannot be used to determine the beginning and the end of the $\alpha \rightarrow B$ or the $\gamma \rightarrow M$ reaction. These reactions are identified by observing changes in the material's properties such as its electrical resistivity, specific volume, or magnetism.

An isothermal transformation diagram is constructed by plotting on temperature-log time coordinates the point at each temperature where the austenite begins to transform and the point where the transformation is complete. Two heavy lines are then drawn through the two sets of points, the left one indicating the beginning of transformation and the right one designating the end of transformation. Then the points corresponding to 50% transformation are plotted and a light dashed line is drawn through them. A TTT diagram constructed in such a fashion is shown in Figure 5-6. Diagrams such as this exist for all of the common alloy steels.

In addition to knowing how an isothermal transformation diagram is constructed, it is important to know how to "read" or interpret the diagram. The following examples of thermal treatments with AISI 1080 steel and Figure 5-6 will help the reader master their use.

Consider a specimen that is austenitized at 1400°F and then cooled to 1000°F in less than one second. At this point on the IT diagram, the austenite is just beginning to transform to ferrite and cementite. Although in this case the ferrite and cementite would have the overall composition of the eutectoid pearlite, this condition is not true in all cases, but only when the steel is of eutectoid composition. This is one of the reasons for not labeling the F + C regions with the symbol P. If the specimen is held at 1000°F for 10 seconds, all of the austenite will have transformed to ferrite and cementite, or fine pearlite. Holding this fine pearlite at 1000°F for somewhat longer periods of time will have no effect on its structure since grain growth is very slow at such a low temperature for steel. However, if the transformation has occurred at 1300°F and the specimen were kept at this higher temperature for many hours, the coarse lamellar pearlite would probably change to a spheroidized pearlite.

If the specimen that has been held at 1000°F for 10 seconds were then water quenched there would be no change other than the slight volume contraction associated with a decrease in temperature. In other words, the microconstituent at 1000°F is fine pearlite and the microconstituent at room temperature is also fine pearlite even though a vertical line drawn down from the (1000°F, 10 sec) point crosses the $\gamma \rightarrow B$ lines. It must be appreciated that once austenite transforms to any one of its transformation products there are no further changes in microconstituents on further cooling, although there may be some changes on reheating.

Consider another specimen that has been cooled instantaneously from 1400°F to 500°F. (In reality this would be very difficult to do.) The material would remain austenite for approximately one minute before it started to change to bainite (F + C), and approximately one hour would be required for the transformation to be complete.

The transformation from austenite to martensite cannot proceed to completion at a constant temperature, as explained previously and shown in Figure 5-6. If a specimen of AISI 1080 steel is cooled at a rate such that the temperature drops from 1400°F to 70°F in one second, the transformation to martensite starts at a temperature of about 430°F and is completed when the temperature reaches 230°F. Reheating martensite to temperatures above room temperature but less than the critical temperature is called tempering and is discussed under the heading "Tempering" on p. 141.

Although the isothermal transformation (IT) diagram for each steel is different from all the others, that is, no two diagrams are exactly alike, IT diagrams do fall into three broad types. One type is the simple eutectoid type illustrated in Figure 5-6. The second contains a proeutectoid phase or microconstituent as illustrated in Figure 5-8. The third type, which occurs with some alloy steels, contains a well defined bainite nose as well as a pearlite nose as illustrated in Figure 5-9.

The diagram of an AISI 1040 steel illustrated in Figure 5-8 is typical of hypoeutectoid and hypereutectoid steels. This type of diagram is different from the eutectoid type shown in Figure 5-6 in that there is a two-phase region above the pearlite nose that separates the single-phase A region from the three-phase A + F + C region. This two-phase region of the IT diagram corresponds to the two-phase regions that lie between the upper and lower critical temperatures of the equilibrium phase diagram. The proeutectoid phase for the hypoeutectoid

Non-Equilibrium Diagrams and Microconstituents 129

steels is ferrite. Thus, the Fe-C equilibrium phase diagram (see Figure 3.10, p. 86) indicates that when a 1040 steel is slowly cooled from the austenite region, or about 1550°F, some ferrite will begin to precipitate out of the austenite when the temperature drops to about 1475°F. And when the temperature reaches 1334°F, the 1040 steel will consist of equal amounts of the two phases austenite and ferrite. Upon very slow cooling past 1333°F the remaining austenite transforms to coarse pearlite.

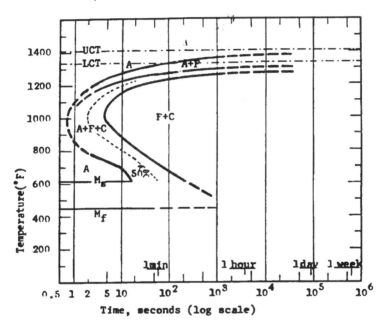

Figure 5-8. Isothermal transformation diagram for an AISI 1040 steel showing a proeutectoid phase. UCT = upper critical temperature; LCT = lower critical temperature.

On the other hand, the IT diagram in Figure 5-8 indicates that if the 1040 steel has been very rapidly cooled from 1550°F to 1300°F, no ferrite would precipitate until after 20 seconds has elapsed and no pearlite $(\alpha + \theta)$ would form during the first hour at 1300°F.

The IT diagram for any hypereutectoid steel is similar to that in Figure 5-8 except that the proeutectoid phase in the two-phase region would be cementite instead of ferrite. The pearlite nose is displaced to the right with increasing carbon or alloy content. This indicates an increase in hardenability, as will be discussed later. The martensite transformation temperatures are lowered as the carbon content increases.

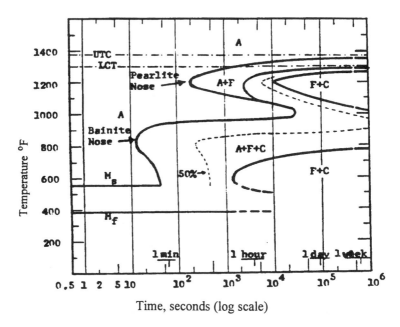

Figure 5-9. Isothermal transformation diagram for an AISI 4340 showing a bainite nose.

The IT diagram for an AISI 4340 steel illustrated in Figure 5-9 contains a bainite nose as well as a pearlite nose and is typical of the third type of IT diagram. This particular steel has a eutectoid carbon content of 0.50% since it has the following composition: 1.80 Ni, 0.80 Cr, 0.33 Mo, and 0.80 Mn. But, since this particular heat has a carbon content of 0.42%, it is a hypoeutectoid steel and therefore it has a proeutectoid ferrite region in front of the pearlite nose. There are three characteristic features of this type of IT diagram: (1) the pearlite nose occurs in the temperature range of 1100°F to 1200°F; (2) the bainite nose appears in the temperature range of 800°F to 900°F; and (3) the bainite nose lies to the left of the pearlite nose.

If a piece of 4340 steel is isothermally transformed in the temperature range 1300°F to 1000°F, the resulting microstructure is pearlite. If a piece is isothermally transformed in the temperature range of 1000°F to 550°F, the microstructure is bainite. If a piece is allowed to cool continuously at such a rate that it cools from about 1550°F to 1200°F in less than 2 minutes and to 600°F in 1 to 2 hours, the resulting microstructure will be bainite. In order to obtain martensite, this steel must be given a continuous cooling rate such that its temperature drops past 800°F (the bainite nose) in less than 12 seconds.

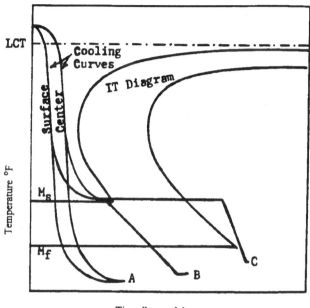

Figure 5-10. Cooling curves superimposed on an IT diagram. (A) Common continuous quench showing that the surface transforms to 100% M before the center starts to transform. (B) The martempering process allows the surface and center to transform to martensite simultaneously. (C) The austempering process allows the surface and center to transform to lower bainite simultaneously.

Interrupted quenching is used very little in the production heat treating of steel because it is not as convenient or economical as the continuous cooling type of quenching treatments. Two special types of interrupted quenching are sometimes used in order to avoid quenching cracks. They are illustrated in Figure 5-10 along with the common continuous cooling quench. Curves A, representing the continuous quench, indicate why quench cracks sometimes appear during heat treating. If the surface of the part being quenched transforms to brittle primary martensite and cools to room temperature while the center is still austenite, then, when the center does transform to martensite, the resulting volume expansion (see Figure 3.9, p. 85) of the interior material causes the surface to be stressed in tension and sometimes to crack.

In the *martempering* process indicated by curves B, the part is quenched into a liquid bath that is held just above Ms temperature for a sufficient time to allow the center to reach the same temperature as the surface, then the piece is withdrawn and air cooled to room temperature. The process permits the surface and center to transform to martensite simultaneously. The steel part must be tempered after this process, just as it must be after the common continuous quench.

The *austempering* process indicated by curves C is one in which the part is quenched into a liquid bath that is held just above the M_s temperature for a sufficiently long time to enable the material to completely transform to bainite. Tempering is not necessary after this process. Lower bainite is slightly softer than martensite tempered at the same temperature.

Even though the IT diagrams are obtained by isothermally transforming the steels, their main application is in predicting the microstructures that result after continuous cooling, and thus they assist in selecting the optimal steel and treatment for a given part. As would be expected, the transformation of austenite to ferrite plus cementite on continuous cooling does not begin at the time and temperature that correspond to the intersection of the cooling curve and the left isothermal transformation curve because the steel is not at that low temperature for the entire time. The transformation actually begins later and at a lower temperature. But once the transformation begins it may continue to completion at a constant temperature since the transformation is an exothermic reaction.

An actual cooling curve that was obtained experimentally at the enter of a 3/4" diameter bar of AISI 1080 steel during air cooling is shown in Figure 5-11. The curve indicates that the austenite began to transform to pearlite (medium) at a temperature of 1240°F approximately 30 seconds after its temperature dropped below the lower critical temperature. The temperature remained at 1240°F for an additional 90 seconds before it started to drop again.

Non-Equilibrium Diagrams and Microconstituents   133

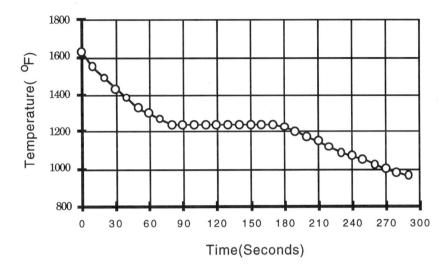

Figure 5-11. Cooling curve at the center of a 3/4" diameter bar of AISI 1080 steel during air cooling.

The above cooling curve is superimposed on an IT diagram for 1080 steel in Figure 5-12. From this plot it is evident that even on a continuous quench the transformation of austenite to pearlite occurs, for the most part, at a constant temperature that depends upon the cooling rate used.

The isothermal transformation diagram for any steel can be used to predict the microstructures that will most likely result from a specified thermal cycle or conversely to determine what cooling rate is necessary to obtain a desired microstructure by modifying the IT diagram to a CT (continuous transformation) diagram. This modification is as simple to make as that required to change an equilibrium phase diagram to an equilibrium microconstituent diagram in that the change usually can be visualized without actually constructing any additional lines.

134   Chapter 5

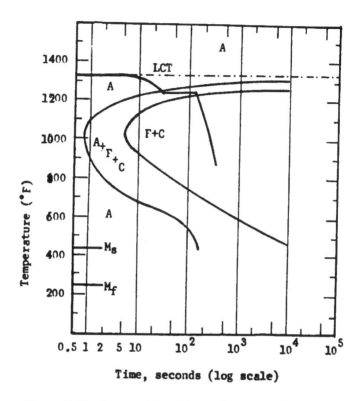

Figure 5-12. Superposition of a cooling curve for air cooled 1080 steel on its TTT diagram.

The relative position of a CT diagram with respect to its corresponding IT diagram is shown in Figure 5-13. The transformation-begins curve of the CT is approximately one third of the way between the transformation-begins and -ends curves of the IT diagram. The transformation-ends curve of the CT diagram lies an equal amount to the right of the corresponding IT curve.

The two cooling rate curves in Figure 5-13 help explain the *critical cooling rate,* which is the slowest cooling rate that steel can undergo and still transform to 100% martensite. The critical cooling rate is also defined as the cooling rate that is tangent to the nose of the IT curve for that particular steel, curve A in Figure 5-13. However curve A is not really the slowest cooling rate that results in 100% martensite; instead the critical cooling rate is defined by curve B which is tangent to the CT diagram.

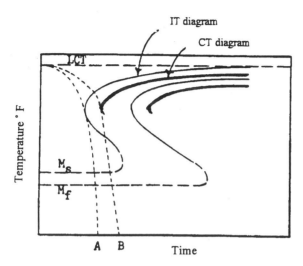

Figure 5-13. Super position of CT and IT diagrams.

It is possible to make reasonable predictions of what microstructures will result in a given steel by superimposing the appropriate continuous cooling rate curve on the IT diagram for that steel, keeping in mind the relative position of the CT curves. This is illustrated in Figure 5-14 for AISI 1080, steel with cooling rate curves characteristic of the cooling rate at 1300°F when pieces less than 1/2" diameter are quenched. The curves plotted here are cooling *rate* curves, not cooling curves, which have a decreasing rate with decreasing temperature. Thus curve (1) has a temperature drop from 1333°F to 1033°F in the first second and a drop to 733°F in the second second. Since the cooling rate for a given type of quench does not vary much in the significant temperature range of 1000°F to 1300°F, the use of the easy-to-construct cooling rate curves is recommended.

Although curve (1) for a water quench intersects the IT transformation-begins curve, it does not intersect the CT. Therefore, the resulting microstructure is all martensite. Curve (2), typical of an oil quench, intersects the left CT curve at a temperature of approximately 1100°F. This results in a final room temperature microstructure of fine pearlite. In similar fashion it can be determined that air cooling induces transformation at about 1250°F, and furnace cooling transforms steel at about 1300°F. The former results in medium pearlite and the latter in coarse pearlite.

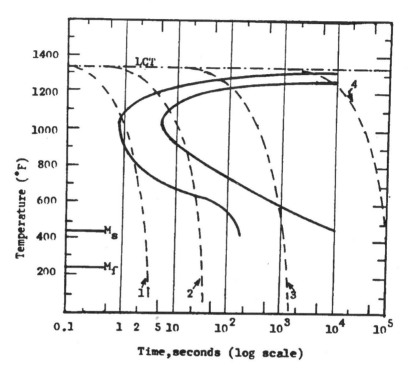

Figure 5-14. Continuous cooling rate curves superimposed on the IT diagram for AISI 1080 steel. (1) 300°F/sec = water quench. (2) 30°F/sec = oil quench. (3) 1°F/sec = air quench. (4) 0.01°F/sec = furnace cool. Small pieces, less than 1/2" diameter in all cases.

HARDENABILITY

Hardenability is a measure of how slow a cooling rate can be used for a given steel to achieve all martensite. It is frequently referred to as the ease with which martensite can be formed. *Hardenability* should not be confused with *hardness*, which is discussed in Chapter 6. At this point it is sufficient to realize that hardness is a measure of a material's compressive strength (after a certain amount of deformation imparted by the indenter), whereas hardenability is a measure of a steel's ability to achieve a high hardness due to the presence of martensite in the center of a large bar or thick section. Consider two steels, A having 0.2% C and B having 0.6% C, where 2" diameter bars are austenitized and water quenched. Bar A has a surface hardness of 45 R and a center hardness of 44 Rc, while bar B has a surface hardness of 65 Rc and a center hardness of 50 Rc. In this case steel A has a higher hardenability than steel B, even though its actual hardness is lower. The hardness of a 0.2 C steel containing all martensite is 45 Rc and that of a 0.6% C steel is 65 Rc. Thus steel A transforms to martensite at the center but steel B does not. In summary, the hardness of steel is determined by the carbon content, whereas the hardenability is determined by the amount of alloys present.

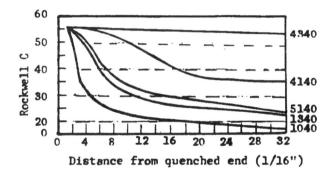

Figure 5-15. Jominy end-quench curves for some medium carbon steels.

The most convenient form of presenting the experimental hardenability data for steel is the end-quench hardness data obtained from a Jominy test because it is much easier to interpret this type of hardness data than an IT diagram. End-quench hardness data is displayed in two forms: (1) in graphical form called Jominy curves, as illustrated in Figure 5-15 and (2) in tabular form, as presented in Tables 6-3 to 6-5. The latter form is preferable because the data are more compact, that is, more data can be presented on one page.

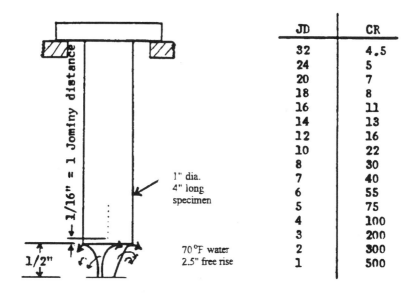

Figure 5-16. Jominy end-quench test and associated cooling rates. JD = Jominy distance (1/16") from quench end. CR = cooling rate (1°F/sec) at 1300°F.

The apparatus used in conducting a Jominy end-quench test is shown in Figure 5-16. Water at 70°F flows out of a vertical piece of 1/2" pipe under a pressure such that it has a free rise of 2 1/2". The standard specimen is 1" diameter and 4" long with provisions for attaching a large ring or washer opposite the end to be quenched to support the specimen in a fixture during the quench. The specimen and ring are austenitized (usually at 1550°F) for 1 hour and then placed in the quenching fixture. The table in Figure 5-16 lists the cooling rate at various locations in the lower 2" of the specimen. The Jominy end-quench test assumes that the thermal conductivities of all the carbon and alloy steels are approximately the same, and so they all have the same cooling rate at a given distance from the quenched end. Thus all steels have a cooling rate of 30°F/sec at 1300°F at 8 Jominy distances (1/2") from the quenched end. The tabular data in Figure 5-16 fit the expression

$$R = 11\,d^{-1.33} \quad \text{for} \quad 0.1 < d < 2.4, \quad \text{or} \quad d = 6.07\,R^{-0.752} \quad (5\text{-}6)$$

Non-Equilibrium Diagrams and Microconstituents 139

where R is the cooling rate in °F/sec at 1300°F and d is the distance in inches from the quenched end.

After the entire specimen has cooled to room temperature, two parallel flats are ground along the entire length and Rockwell C hardness readings are taken at 1/16" intervals starting 1/16" from the quenched end. The average of the readings taken on the two sides is recorded either in tabular form or as a curve of the sort shown in Figure 5-15.

End-quench tests performed on different heats of the same class of steel result in a range of hardness distributions because of the variation in the resulting chemical composition. Steels that are selected on the basis of hardenability calculations should be specified on the basis of guaranteed upper and lower limits to the end-quench curves. Such steels are designated as H steels (the letter H follows the AISI number). Thus an AISI 1340 steel has a guaranteed chemical composition, but no guarantee of its hardenability. On the other hand, an AISI 1340 H steel guarantees both the composition and the hardenability. The guaranteed hardness limits of the H steels are referred to as hardenability bands. The hardenability band for AISI 1340 H steel is shown in Figure 5-17.

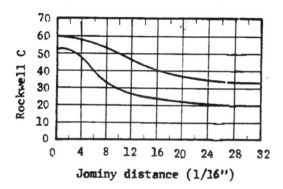

Figure 5-17. Hardenability band for AISI 1340 H steel.

End quench hardenability data is very useful in the selection and specification of steel because *the cooling rate for a given steel determines the resulting microstructure*. Therefore, regardless of how a specific cooling rate is achieved, the resulting microstructure is the same. Thus the center of a small bar quenched in oil can have the same cooling rate at 1300°F as a larger bar quenched in water, and both will have the same microstructure and strength. However, the microstructure at the surface of these two bars will be different.

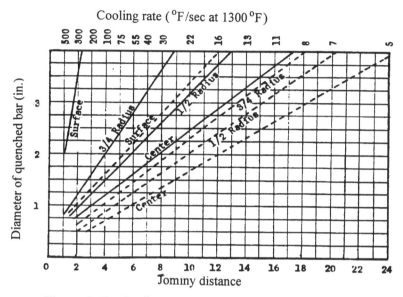

Figure 5-18. Cooling rates and equivalent Jominy distances of round quenched bars. H is the severity of quench. Solid lines: quenched in still water, moderately agitated; H = 1.5. Dashed lines: quenched in still oil, moderately agitated; H = 0.5.

In order to use end-quench hardenability data to select a steel for a given part, it is necessary to determine the cooling rates at the critical locations in the part. Theoretically this can be done by means of heat transfer calculations, but some practical difficulties are encountered due to the absence of surface heat transfer coefficients of materials having varying amounts of heat-treating scale. The most common method for determining the cooling rates of quenched parts is to use charts or tables as are found in various handbooks for the cooling rates of simple shapes such as cylinders, spheres, and plates. Figure 5-18, an example of such a chart, gives the cooling rate (and equivalent Jominy distance) at various locations in round bars quenched in either water or oil. In order for these curves to be valid, the ratio of the length to the diameter must be

Non-Equilibrium Diagrams and Microconstituents 141

greater than 3 and the positions along the length of the bar must not be closer than one diameter from the end of the bar.

The microstructures and mechanical properties of a given steel can be very easily determined by means of cooling rate charts and end-quench hardenability data. For example, consider a 2" diameter shaft made of AISI 1340 H steel that is water quenched. From Figure 5-18 the cooling rate at the surface is found to be 500°F/sec (1 JD) and at the center it is 30°F/sec (8 JD). From the hardenability band in Figure 5-17, the hardness at the surface is found to vary between a maximum of 60 Rc (653 H_B) and a minimum of 52 Rc (495 H_B). If the lower value is used in the design of the shaft, no factor of safety need be applied to account for material variations. Likewise, the hardness at the center can vary from 53 Rc to 33 Rc. If the shaft is simply transmitting torque, then the stresses at the center are low and the hardness need not be high. However, if the 2" diameter bar is carrying a uniaxial load, then the stresses at the center and surface are the same and the low hardness at the center is critical. Examples of the use of hardenability data are presented in Chapter 6.

TEMPERING

Tempering is the process of reheating primary martensite to temperatures below the lower critical temperature for the purpose of removing residual quenching stresses or increasing the toughness of the steel. All structural parts made of heat-treated steel should have a tempering treatment included as part of their treatment to remove the residual quenching stresses since the presence of these stresses cannot be easily ascertained. The term is also applied to (1) the reheating of fine pearlite and (2) a specified amount of cold work, particularly for nonferrous materials. Thus spring temper means that a material has been given 60.5% cold work.

Figure 5-19 illustrates the effect of the tempering temperature on the hardness and strength of three steels: AISI 1080, 1040, and 1000. Other steels are similarly affected. It should be noted that the tensile strength scale in ksi is 0.5 times the Brinell hardness scale. Thus a Brinell hardness of 400 is equal to a tensile strength of 200 ksi. This relationship is discussed in Chapter 6, along with several other important characteristics of the curves in Figure 5-19 that are common to all tempering charts.

142   Chapter 5

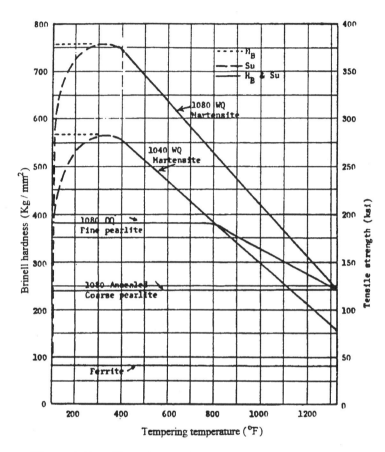

Figure 5-19. Effect of tempering on the hardness and strength of steel.

The tensile strength of primary as-quenched martensite may be very low even though the hardness is high; thus the above-mentioned 2-to-1 ratio does not always apply. For example, 0.8% C martensite has a Brinell hardness of 760 (67 Rc), but its tensile strength may vary from less than 1 ksi to about 350 ksi. Likewise, a 0.4% C martensite steel will have a hardness of 570 HB (57 Rc), but its tensile strength may vary from nearly zero to 280 ksi. The reason for this variation in the tensile strength of primary martensite is not its tetragonal lattice nor its being supersaturated with carbon but simply the

Non-Equilibrium Diagrams and Microconstituents   143

presence of macroresidual tensile stresses at the surface of the bar. These tensile stresses are caused by two conditions: (1) the surface of the bar transforms to martensite before the center does, and (2) the specific volume of martensite is greater than that of the austenite. Although the latter is a constant for a given steel, the former is influenced by the size and shape of the part and by the quenchant used. Occasionally the residual stresses are so large that the part cracks during the final cooling of the heat-treating process even before any loads are applied to it.

By tempering the primary martensite at a temperature of approximately 300°F, the steel achieves its maximum strength without any appreciable reduction in hardness. However, its impact strength is reduced at this high hardness level.

As the tempering temperature is increased, the hardness and strength decrease approximately linearly until the lower critical temperature is reached. Steel tempered just below the critical temperature has mechanical properties that are approximately the same as in the annealed condition. Since strength is nearly a linear function of the tempering temperature, it is very easy to interpolate the strength, or hardness, of a steel after any tempering treatment. This interpolation is made even simpler by assuming that the steel has its as-quenched hardness reduced by 5% when tempered at 330°F and that it has its annealed hardness when tempered at 1330°F. The slope (which is negative) of the line, when this function is graphed, is the difference between the as-quenched and annealed hardnesses divided by 1000. The following relationships between the strength or hardness after tempering and the tempering temperature are as follows:

$$(H_B)_x = \frac{0.95(H_B)_Q - (H_B)_A}{1000}[1330 - T_x] + (H_B)_A \qquad (5\text{-}7)$$

$$(S_u)_x = \frac{(S_u)_Q - (S_u)_A}{1000}[1330 - T_x] + (S_u)_A \qquad (5\text{-}8)$$

In these equations Q and A stand for as-quenched and annealed, respectively. $(S_u)_Q$ must be the stress-relieved value and not the actual as-quenched value obtained on a specimen with residual stresses. Actually, the tensile strength after a 330°F temper is the most appropriate value to use. The following example will make the reader more familiar with these relationships.

Consider a 1080 steel water quenched to form martensite. The as-quenched hardness is about 760 H_B and the annealed hardness is 240 H_B, so Eq. 5-7 can be rewritten as

$$(H_B)_x = 0.48(1330 - Tx) + 240 \qquad (5-9)$$

to give the hardness after tempering at any temperature Tx. Thus the Brinell hardness after tempering at 900°F is

$$H_B = 0.48 \times 430 + 240 = 446$$

The tensile strength, in ksi, is 0.5×446 or 223.

These relationships will be developed further in Chapter 6. The best source of data for mechanical properties of steel after tempering is Bethlehem Steel's handbook *Modern Steels and Their Properties*. However, care must be exercised in using the data in this handbook because in the case of the plain carbon steels the hardness was measured at the surface of a 1" diameter quenched bar and the tensile strength was measured near the center of the bar. Also, caution must be exercised in comparing a plain carbon steel to any alloy steel because in the former case a 1" diameter bar was quenched and in the latter case a 0.532" bar diameter was quenched. In both cases a 0.505" diameter specimen was machined and tested.

The curve for ferrite is included in Figure 5-19 as a reference or guideline. Steel having the carbon content of ferrite does not transform to martensite even though it does transform to austenite above 1680°F. The hardness of ferrite does not change by heat treatment, but it is affected by grain size and trace amounts of impurities. The value of 80 H_B is typical for ferrite having an average grain size and may be slightly higher or lower as the grain size decreases or increases, respectively.

Figure 5-20 illustrates qualitatively the trade-off in strength that occurs in order to increase the toughness of a metal. This is true of the nonferrous materials as well as the steels. Heat treating to maximize a material's tensile or yield strength will have minimized its ductility or toughness, and vice versa. This figure illustrates two features of tempering as it applies to steel. First, most steels show a reduction in impact strength if tempered at 500°F to 600°F with no corresponding change in the other mechanical properties. Since steel that is heated to this temperature range in air takes on a blue color owing to the oxide formed, the phenomenon of the reduction in toughness associated with this tempering temperature is called *blue brittleness*. Second, metallurgists formerly referred to an "optimal combination" of properties resulting from tempering the quenched steel to the temperature at which the strength dropped to about half its value and the toughness approximately doubled. This practice is not followed as much today because the design engineer usually determines exactly what properties are needed and then specifies the treatment that will provide these properties.

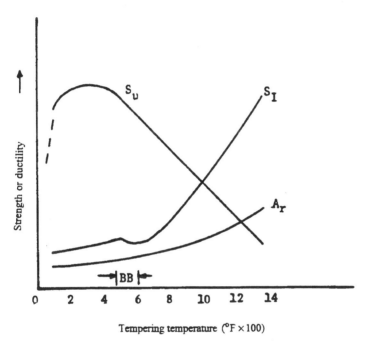

Figure 5-20. Effect of tempering temperature on the mechanical properties of steel.

Tables 5-1 and 5-2 illustrate the effect of tempering on the mechanical properties of AISI 1040 water-quenched steel and 4140 oil-quenched steel. These data are typical of data presented in many mechanical design textbooks. Too frequently only a superficial study is made of tempering data such as these, and the reader gets an exaggerated impression of the relative properties of plain-carbon steel compared with alloy steels. For example, it is readily apparent from Table 5-1 that the maximum tensile strength obtainable with a 1040 water-quenched steel is 130,000 psi whereas the same carbon content in an alloy steel, such as the 4140 shown in Table 5-2, produces a maximum strength of 290,000 psi.

Table 5-1. Effect of Tempering on the Mechanical Properties of AISI 1040 Steel.*
(Avg. values. Size treated: 1" rd.; Size tested: 0.505" rd.)
(Normalized at 1650°F, reheated to 1550°F, water quenched.)

Tempering Temp. (°F)	Yield Strength (ksi)	Tensile Strength (ksi)	Elongation (%)	Area Reduction %	Brinell Hardness
400	97	130	19	46	514
600	94	129	19	52	444
800	92	123	21	57	352
1000	87	114	23	62	269
1200	72	97	27	68	201
1300	65	87	33	69	

*Data from Bethlehem Steel Company's handbook *Modern Steels and Their Properties*.

Table 5-2. Effect of Tempering on the Mechanical Properties of AISI 4140 steel.*
(Single heat values. Size treated: 0.530" rd; size tested: 0.505" rd.)
(Normalized at 1600°F. reheated to 1550°F, oil quenched.)

Tempering Temp. (°F)	Yield Strength (ksi)	Tensile Strength (ksi)	Elongation %	Area Reduction (%)	Brinell Hardness
400	254	290	10	42	578
600	230	252	12	45	495
800	195	209	14	48	429
1000	153	167	16	52	341
1200	115	130	20	59	277
1300	100	116	23	63	235

*Data from Bethlehem Steel Company's handbook *Modern Steels and Their Properties*.

It is apparent from these tables that both the size of the specimen and the agitation of the quenching fluid are different. In the case of the alloy steel, a 0.530 in. diameter specimen was quenched in agitated oil and then a 0.505 in. diameter specimen was tested. In the case of the plain carbon steel, a 1 in. diameter specimen was quenched in water and then a 0.505 in. diameter specimen was tested. For the sizes quenched, the alloy steel formed martensite to the center whereas the larger plain-carbon steel did not form martensite to the center. However, if the sizes of the pieces undergoing heat treatment were reversed, what would their tensile strength be? The answer lies in the hardenability data.

Figures 5-21 and 5-22 compare the effects of reheating a solution-treated aluminum alloy and an iron alloy in terms of both microconstituents and hardness.

For the aluminum alloy this process is called precipitation hardening since the microstructure containing the fine dispersion of the $Cu-Al_2$ particles in the alpha matrix is much harder and stronger than the supersaturated solid solution that results from the rapid quench from the high temperature.

In the case of the iron alloy (steel), the result of the reheating after the rapid cooling from the high temperature is a softening of the metal. This is because the high-temperature phase (face-centered cubic austenite) transforms to the nongranular acicular solid solution (martensite) by a shear process. This supersaturated solid solution of carbon in iron is extremely hard, about 67 R_C or 800 H_B. When the carbon precipitates out as iron carbide, the hardness decreases to about 180–240 H_B.

148   Chapter 5

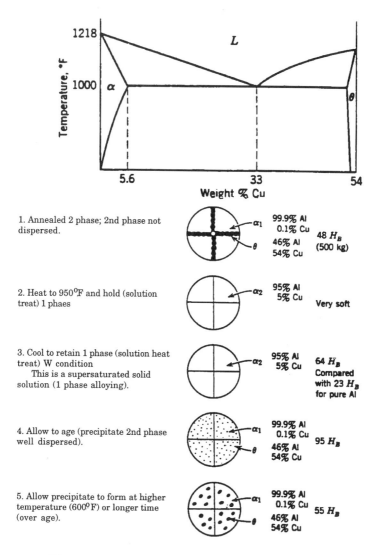

Figure 5-21. The heat treatment, microstructures, and hardness of a 95% Al-5% Cu alloy.

Non-Equilibrium Diagrams and Microconstituents 149

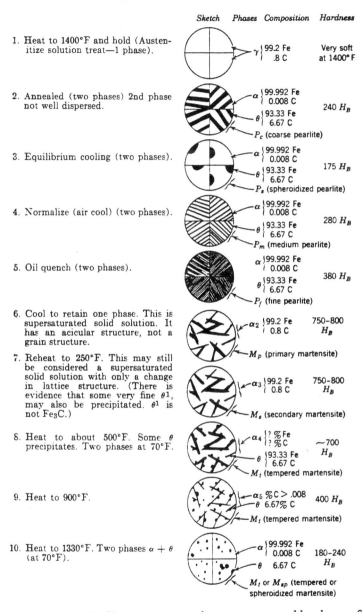

1. Heat to 1400°F and hold (Austenitize solution treat—1 phase).

2. Annealed (two phases) 2nd phase not well dispersed.

3. Equilibrium cooling (two phases).

4. Normalize (air cool) (two phases).

5. Oil quench (two phases).

6. Cool to retain one phase. This is supersaturated solid solution. It has an acicular structure, not a grain structure.

7. Reheat to 250°F. This may still be considered a supersaturated solid solution with only a change in lattice structure. (There is evidence that some very fine θ^1, may also be precipitated. θ^1 is not Fe_3C.)

8. Heat to about 500°F. Some θ precipitates. Two phases at 70°F.

9. Heat to 900°F.

10. Heat to 1330°F. Two phases $\alpha + \theta$ (at 70°F).

Figure 5-22. Heat treatment, microstructures, and hardness of a 99.2% Fe-0.8% C alloy.

150 Chapter 5

EXAMPLE PROBLEMS

5-1. Why are hypoeutectoid steels heated above the upper critical temperature and hypereutectoid steels heated to slightly above the lower critical temperature in the hardening operation?

Solution: With the hypoeutectoid steels, the object is to obtain the maximum strength which is accomplished by achieving 100% stress-relieved martensite. With hypereutectoid steels, the object is to obtain a wear-resistant material as well as a material with strength. Therefore, by raising the temperature slightly above the lower critical temperature, we obtain carbides dispersed in a hard martensitic matrix after quenching.

5-2. Four small pieces of 1040 steel are heated to 1350°F and cooled as indicated below. Sketch and label the microstructures and give the hardness of each: (a) furnace cool, (b) air cool, (c) oil quench, and (d) water quench.

Solution:

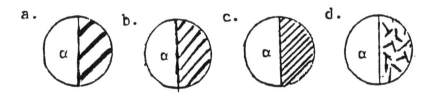

1/2 α, 1/2 P_c	1/2 α, 1/2 P_m	1/2 α, 1/2 P_f	1/2 α, 1/2 M
1/2 × 80 +	1/2 × 80 +	1/2 × 80 +	1/2 × 80 +
1/2 × 240	1/2 × 280	1/2 × 380	1/2 × 740
H_B = 160	180	230	410

5-3. What is the function of carbon as an alloy in steel? Of the alloys Mn, Cr, Ni, Mo?

Solution: Carbon is the element that imparts the hardness to martensite up to a maximum of 0.65%. The other principal alloying elements are added to increase the hardenability of the steel–the ability to form martensite with a lower cooling rate.

Some elements that do not go into solution in the iron lattice are added to improve the machinability of the steel. Lead is such an element. Also, larger amounts of manganese and sulfur together form an insoluble MnS inclusion. In all such cases, the improved machinability is principally due to a large reduction in the transverse fracture strain of the rolled or drawn bars of the metal.

REFERENCES

1. Van Vlack, L.H. *Elements of Materials Science*, 2nd ed. Addison Wesley, Reading, MA, 1989.
2. Flin, R.A. and P.K. Trojan, *Engineering Materials and Their Application,* Houghton Mifflin, Boston, 1990.
3. United States Steel, *Atlas of Isothermal Transformation Diagrams*, United States Steel Co., Pennsylvania, 1951.
4. Datsko, J. *Material Properties and Manufacturing Processes*. John Wiley & Sons, New York, 1967.
5. Hollomon, J.H. and L.D. Jaffe, *Ferrous Metallurgical Design*. John Wiley & Sons, New York, 1947.
6. Craft, W., and J.L. Lamont, *Hardenability and Steel Selection*. Pitman Publishing, New York, 1949.
7. Bethlehem Steel Co. *Modern Steels and Their Properties,* Bethlehem Steel Co., Bethlehem, PA, 1972.

Non-Equilibrium Diagrams and Microconstituents 153

STUDY PROBLEMS

5-1. From the data below, illustrate (including sketches of microstructures):

    a. The magnitude of the effect of cold work alone on pure aluminum.
    b. The magnitude of the effect of solution hardening alone on aluminum.
    c. The magnitude of the effect of precipitation hardening alone on aluminum.

Material	S_y	S_u	H_B
1100-O	5,000	13,000	23
1100-F	14,000	16,000	28
1100-H18	22,000	24,000	44
2011-O	15,000	30,000	48
2011-W	18,000	36,000	63
2011-T3	48,000	55,000	95
2011-T4	45,000	50,000	90

5-2. What is the strength-to-density ratio for steel? For aluminum?

5-3. What are the maximum tensile strength, yield strength, and fatigue strength that can be achieved in a 1/2"–13NC threaded stud of the following materials: *(Include your reference or source.)*

    a. 1040 steel
    b. 4140 steel
    c. 1020 steel

5-4. What maximum tensile load can a 1/2"–13 NC stud support if it is made of the following: *Note:* γ'd means *austenitized.* Area = 01416 in^2.

    a. 1020 annealed steel
    b. 1020 steel = γ'd 1650°F, iced brine quench, T 400° F
    c. 1040 steel = γ'd 1550°F, W.Q., T 400° F
    d. 4140 steel = γ'd 1550°F, O.Q., T 400° F

154   Chapter 5

5-5. Two shafts of 1/2 in. diameter are made of AISI 1040 steel and heat treated as specified below. What are the percentages of each of the microconstituents present and what are the probable tensile and yield strengths? *Note:* γ'd means *austenitized.*

    a. γ'd 1400°F; water quenched and tempered at 450°F
    b. γ'd 1550°F; water quenched and tempered at 450°F

5-6. A certain steel has the following composition: C = 0.40%; Mn = 1.00%; Cr = 0.50%; Si = 0.25%. What is the amount of each microconstituent after the steel is annealed? Calculate the values of its hardness, yield strength, and tensile strength.

5-7. What are hypoeutectoid steels used for? What are hypereutectoid steels used for?

5-8. If 1020 annealed steel has an A_r of 66 and 1040 steel has an A_r of 58, what is the percent reduction of area of ferrite and of coarse pearlite?

5-9. Calculate the properties of the following materials on the basis of their microstructure (small pieces).

    a. Brinell hardness number (Bhn) of 1060 steel—annealed.
    b. Tensile strength of 1030 steel—small piece, oil quenched from 1550°F.
    c. Yield strength of 1010 steel water quenched from 1340°F.

5-10. A bronze of 85% Cu–15% Sn is a good material for bearings, whereas a brass of 85% Cu–15% Zn is not. On the basis of the equilibrium diagrams, explain why.

5-11. List all of the practical ways the following materials may be strengthened without changing the composition:

    a. 70% Cu–30% Zn brass
    b. 5% Cu–95% Al alloy
    c. 1% C–99% Fe alloy
    d. pure magnesium

Chapter 6

THERMAL STRENGTHENING AND COSTS

INTRODUCTION

This chapter presents detailed information and examples to enable the reader to select or specify a material on the basis of its strength after heat treating. Cost information is given so that one can determine the cost-effectiveness of a material and select the lowest cost material on the basis of mechanical strength. Since the ferrous metals are the materials most commonly used in manufacturing, and since their microstructures are more completely understood, they are emphasized. Cost and mechanical property data are included for several non-ferrous metal alloy systems and for the common plastics so that one can compare their cost effectiveness to ferrous metals.

FERROUS METALS

Table 6-1 lists the mechanical properties of all of the microconstituents that are associated with steel, both plain carbon and alloy. These values do not apply to any of the stainless steels or cast irons. The microstructures that result from a given thermal treatment can be predicted by means of the equilibrium and non-equilibrium diagrams discussed in the previous chapters. The following

156   Chapter 6

section describes a procedure for predicting the hardness, and thus indirectly the microstructure, of a heat-treated steel on the basis of its chemical composition.

Table 6-2 lists some very useful relationships, developed by the author during the past 35 years, between the carbon content of steel and the resulting mechanical properties. These relationships are expressed mathematically — a presentation more useful than the traditional qualitative or tabular treatment.

Table 6-1. Mechanical Properties of Ferrous Microconstituents
(Microconstituents vs. cooling rate are for carbon steel)

Micro-constituent	Cooling rate (°F/sec)	H_B^1	S_u^2 ksi	S_u^2 MPa	σ_o ksi	σ_o MPa	ε_u	ε_f	Izod ft.-lb.
α	any	80	40	275	77	534	0.30	1.60	100
Ps	0.01	180	90	620	130	902	0.12	0.80	10
Pc	0.05	240	120	1655	167	1151	0.10	0.50	5
Pm	1	280	140	1930	195	1343	0.10	0.40	5
Pf	30	380	190	2620	252	1737	0.08	0.40	10
M	300-500	300-800	_3	_3	_3	_3	_3	0-0.1	15
Mt		180-700	_3	_3	_3	_3	_3	0-1.2	180
B		300-800	_3	_3	_3	_3	_3		
θ^4		800-1800	_3	_3	_3	_3	0		0

^{1}3000 kg load and 10 mm dia. ball.

$^2 S_u(ksi) = 0.5 \times H_B$; $S_u(MPa) = 3.45 \times H_B$

^{3}Depends upon carbon content and tempering temperature.

^{4}Depends upon crystallographic orientation. Use 800 H_B to correlate with the strength as a microstructure.

TEMPERING

The effects of tempering on the mechanical properties of the microstructures of steel, as shown in Figure 5-19, can be summarized as follows:

Thermal Strengthening and Costs 157

1. None of the mechanical properties of ferrite or coarse pearlite are changed by tempering.
2. The hardness and strength of fine pearlite are unaffected by tempering up to about 800°F but decrease linearly as the tempering temperature is increased from 800°F to the lower critical temperature (1330°F). When tempering occurs at the lower critical temperature, the hardness (and strength) is approximately the same as that of the annealed microstructure. That is, fine pearlite tempered at 1330°F has about the same mechanical properties of coarse pearlite.
3. The Brinell hardness of martensite tempered (stress-relieved) at about 350°F is about 5% lower than the hardness of the primary martensite. The Brinell hardness decreases linearly as the tempering temperature is raised from 350°F to the lower critical temperature. When martensite is tempered at the lower critical temperature (1330°F), the hardness and other mechanical properties are about the same in the annealed condition. Thus it is possible to estimate the hardness of any heat-treated and tempered steel on the basis of the mechanical properties of the individual microconstituents.

END-QUENCH HARDENABILITY LIMITS

Some experimentally determined hardenability limits for a variety of standard AISI plain carbon steels are given in Table 6-3. The 15zz series of steel is a recent development. It has a manganese content lower than the 13zz series and contains no other alloys. These steels are inexpensive and are satisfactory for parts requiring low hardenability, such as shafts subjected to torsion or bars subjected to bending loads. Appendix E interprets the code for the AISI specification of steel on the basis of composition and treatment.

Table 6-4 lists some of the boron steels. Boron is an alloy for moderately increasing the hardenability of steel. Boron steel is inexpensive since so little is used. The maximum amount of boron that can be added to achieve the greatest increase in hardenability is 0.002%. In some cases, as little as 0.0005% of boron is satisfactory for a sufficient increase in hardenability. The cost of adding boron and other elements that are used to increase hardenability is given in Tables 6-10 and 6-11, along with the quantitative relationship between the amount of

Table 6-2. Carbon Content and Mechanical Properties of Steel

$H_B = 80 + 200 \times \%C$ (up to 0.8%C) for annealed plain carbon steel

$H_B = 80 + 1.60 \times \%Pc$ for annealed carbon and alloy steel

$H_B = 900 \ C^{0.45}$ (up to 0.65%C) for primary martensite

H_B of stress-relieved (180 to 200°C) martensite = 0.95 H_B of primary martensite

$R_C = 73 + 17 \ln \%C$ for $0.10 < C < 0.65$

$R_C = 0.845 \ H_B^{0.66}$ for $400 < H_B < 750$

$R_C = 0.025 \ H_B^{1.25}$ for $220 < H_B < 400$

$R_B = 3.85 \ H_B^{0.6}$ for $100 < H_B < 220$

$S_u = 40 + 100 \times \%C$ ksi (up to 0.8%C) for annealed plain carbon steel

$S_u = 276 + 689 \times \%C$ MPa (up to 0.8%C) for annealed plain carbon steel

$S_u = 0.5 \ H_B$ ksi for stress-relieved and non-cold-worked steel

$S_u = 3.45 \ H_B$ MPa for stress-relieved and non-cold-worked steel

$S_y = 1.05 \ S_u - 30$ ksi for stress-relieved and non-cold-worked steel

$S_y = 1.05 \ S_u - 207$ MPa for stress-relieved and non-cold-worked steel

$I_{max} = 120 - 120 \times \%C$ ft-lb.

---

each alloy and its effect on the hardenability. Costs for non-ferrous and non-metallic materials are given in Tables 6-12, 6-13 and 6-14.

Table 6-5 lists the hardenability limits of most of the standard AISI alloy steels. The Rockwell C values for the hardness of the lower limit of the bands are not listed for hardness of less than 20 R_C since such values are unreliable.

Table 6-6 contains data for determining the equivalent Jominy distance (JD) at various radial distances in round bars quenched with a range of severity of quench from H = 0.2 (quenched in warm oil with no agitation) to H = ∞ (the ideal quench discussed in the following section). For example, a 3" diameter

Thermal Strengthening and Costs

Table 6-3. End-Quench Hardenability Limits for Some Unalloyed Steels

AISI No.	1	2	3	4	6	8	10	12	14	16
					JD (1/16")					
1038H	58-51	55-34	49-26	37-23	28-21	26-	25-	24-	23-	21-
1045H	62-55	59-42	52-31	38-28	32-25	30-24	29-22	28-21	27-20	26
1060	64-57	63-57	62-54	61-33	39-33	39-32	39-31	38-30	38-28	37-27
1080	66-58	66-57	66-52	66-39	44-38	43-38	42-37	42-35	41-34	41-33

Table 6-4. End-Quench Hardenability Limits for Some Higher Manganese and Boron Steels

AISI No.	2	4	6	8	10	12	14	16	18	20	22	24	26	28	30	32
								JD (1/16")								
1330H	56-47	53-40	50-31	45-26	42-23	39-21	37-	35-	34-	33-	32-	31-	31-	31-	30-	30-
1340H	60-52	58-49	56-40	54-33	51-29	48-27	44-25	41-24	39-23	38-23	37-22	36-22	35-21	35-21	34-20	34-20
1522	47-32	45-22	39-20	30-												
1541H	59-50	57-44	55-38	48-27	39-23	33-22	32-21	31-20	30-							
15B37H	56-50	54-48	52-37	50-26	45-22	40-21	33-22	39-								
40B37H	58-50	56-38	53-28	50-25	45-23	38-22	32-22	29-21	28-20	27-	27-	26-	26-	25-	24-	24-
50B60H	66-60	66-60	65-53	65-53	64-42	64-37	63-35	62-34	60-33	58-31	55-30	53-29	51-28	49-27	47-26	44-25
94B40H	60-53	59-52	58-51	57-48	57-43	56-37	55-34	54-31	52-29	50-28	48-27	46-26	43-25	41-25	40-24	39-24

Table 6-5 End-Quench Hardenability Limits for Some Alloy Steels

AISI No.	2	4	6	8	10	12	14	16	18	20	22	24	26	28	30	32
2515H	44-37	42-30	40-24	37-20	34-	31-	29-	28-	27-	26-	25-	24-	24-	23-	23-	22-
3120H	47-39	42-30	35-23	30-20	28-	27-	26-	25-	24-	23-	23-	22-	22-	21-	21-	21-
3140H	60-52	59-49	57-45	56-41	54-36	52-33	50-31	48-30	47-26	44-28	43-28	42-27	41-27	40-26	40-26	39-25
3310H	43-36	42-35	42-33	41-31	40-30	40-29	39-28	38-27	37-26	37-26	37-26	36-26	36-25	36-25	35-25	35-25
3316H	47-39	46-38	46-37	45-35	45-33	45-32	44-32	44-31	44-31	43-31	43-31	43-31	42-31	42-30	42-30	41-30
4032H	54-45	46-29	34-23	29-21	26-	25-	24-	23-	23-	22-	22-	21-	21-	20-		
4068H	≥60	≥59	64-45	62-36	58-34	52-33	47-32	44-32	42-31	41-31	40-30	39-30	38-29	38-29	38-28	37-28
4118H	46-36	35-23	28-	25-	23-	21-	20-									
4130H	55-46	51-38	47-31	42-27	38-26	35-25	34-24	33-23	32-22	32-21	32-20	31-	31-	30-	30-	29
4140H	60-53	59-51	58-50	57-47	56-42	55-39	54-37	53-35	52-34	51-33	49-33	48-32	47-32	46-31	45-31	44-30
4150H	65-59	65-58	65-57	64-56	64-55	63-53	62-50	62-47	61-45	60-43	59-41	59-40	58-39	58-38	58-38	58-38
4320H	47-38	43-32	38-27	34-23	31-21	29-20	27-	26-	25-	25-	24-	24-	24-	24-	24-	24-
4340H	60-53	60-53	60-53	60-52	60-52	59-51	58-49	58-48	58-47	57-46	57-45	57-44	57-43	56-42	56-41	56-40
4620H	45-35	39-24	31-	27-	25-	23-	22-	21-	21-	20-						
4640H	60-52	58-50	56-44	53-37	49-32	44-29	41-27	39-27	38-26	37-26	36-26	35-25	35-25	34-25	34-24	33-24
4720H	47-39	39-27	32-21	28-	26-	24-	23-	22-	21-	21-	21-	20				
4820H	48-40	46-38	43-31	40-27	37-25	35-23	33-22	31-21	29-20	28-20	28-	27-	27-	26-	26-	25-
5046H	62-55	56-32	46-27	35-25	33-24	32-23	31-22	30-21	29-20	28-	28-	26-	25-	24-	23-	23-
5130H	55-46	51-39	47-32	42-28	38-25	36-22	34-20	33-	32-	31-	30-	29-	27-	26-	25-	24-
5150H	65-58	63-56	61-49	59-38	56-34	53-32	50-31	47-30	45-29	43-28	42-27	41-26	40-25	39-25	39-23	38-22
5160H	≥60	65-59	64-56	63-47	61-39	59-36	56-35	52-34	48-33	47-32	46-31	45-30	44-29	43-28	43-28	42-27
6120H	47-38	42-29	36-24	33-22	31-21	31-20	30-	29-	28-	28-	27-	26-	25-	25-	24-	23
6150H	65-58	64-56	63-53	61-47	60-41	58-38	55-36	52-35	50-34	48-32	47-31	46-30	45-29	44-27	43-26	42-25
8620H	47-37	41-27	34-21	30-	28-	26-	24-	24-	23-	23-	23-	23-	23-	22-	22-	22
8630H	55-46	52-39	47-32	41-28	37-26	34-24	33-22	31-21	30-21	30-20	29-20	29-	29-	29-	29-	29
8645H	63-56	63-54	61-50	60-45	58-39	55-35	52-33	49-31	47-30	45-29	43-28	42-28	42-27	41-27	41-27	41-27
8660H	≥60	≥60	≥59	≥57	≥53	≥47	≥44	65-42	64-40	64-39	63-38	62-37	62-36	61-36	60-35	60-35
8720H	47-38	42-30	35-24	31-21	29-	27-	26-	25-	24-	24-	24-	23-	23-	23-	22-	22
8750H	65-59	64-57	63-56	62-53	61-49	60-45	59-42	58-39	57-37	55-35	53-34	52-33	51-33	50-32	49-32	48-32
9260H	66-60	64-53	62-41	58-36	52-35	47-34	43-33	40-32	38-31	37-31	36-30	36-30	35-29	35-29	35-28	34-28
9262H	≥60	≥60	65-56	64-48	62-39	59-37	55-36	48-35	45-34	43-33	41-33	39-32	38-31	37-31	36-30	36-30
9850H	65-59	65-59	65-58	65-58	65-58	65-58	64-57	64-56	63-54	62-52	62-50	61-49	61-48	61-47	60-47	60-47

From "Materials in Design Engineering", 56, No. 5, October 1962 bar quenched in still water (H = 1.0) cools at a rate equivalent to 12 JDs at the center (r/R = 0). However, the surface of the 3" bar will cool at a rate equivalent to approximately 2 1/2 JDs. This latter value is interpolated from the table for r/R = 1 that gives 2 JDs for a 2" diameter bar and 3 JDs for a 3.98" diameter bar.

From the data in Table 6-6, it is possible to construct curves of bar diameter as the ordinate and JD as the abscissa with the r/R locations as parameters for any of the severity of quenches H. Such curves make interpolations easier than using only tabular data, and they can be plotted by a computer.

HARDENABILITY AND COMPOSITION

In order to select the optimal steel for a given part, it is necessary to know both the cost of the individual alloying elements as well as the quantitative effect that each element has on hardenability. "Optimal" in this case means the lowest-cost steel that will properly perform the required function (strength) and have good fabricability.

The concept of an "ideal critical diameter"[1] was created as a means to quantitatively evaluate the effect of the individual alloying elements upon the hardenability of steel. The ideal critical diameter, represented as DI, is the diameter of a round bar of steel that will harden with 50% martensite at the center with an "ideal" quench (H = ∞). The severity of the quench is a measure of how fast the heat is extracted from the hot steel. Table 6-7 lists some experimentally determined values of H varying from 0.02 for cooling in still air to 9.0 for spray quenching with brine (salt water solution).

Table 6-8A lists the multiplying factors for the carbon content and grain size of the steel during austenitizing. For a steel having no other alloying elements, these factors are actually the ideal critical diameter. Thus for a coarse-grained steel (GS #2) having 0.40% carbon, the multiplying factor $fc = 0.41 \times 0.4 + 0.15 = 0.31$; and the ideal critical diameter is 0.31 inches. In other words, a 5/16" diameter bar of this steel when austenitized and cooled with an infinite severity of quench will have 50% martensite at the center. Likewise, a plain carbon coarse-grained steel having 1% carbon has an ideal critical diameter of 9/16."

---

[1]. Grossman, M.A. "Principles of Heat Treatment", ASM 1935, p. 80.

Table 6-6. Bar Diameter (in.) for Equivalent Jominy Distances (JD) with Various Severity of Quenches

	r/R =0 (center)						r/R = 0.3				
JD	H=0.2	0.5	1.0	2.0	5.0	∞	H=0.2	0.5	1.0	2.0	5.0	∞
1	0.15	0.30	0.40	0.70	0.80	1.00	0.10	0.25	0.50	0.65	0.80	1.00
2	0.20	0.45	0.70	0.90	1.09	1.30	0.20	0.50	0.70	1.00	1.20	1.20
3	0.20	0.60	1.00	1.20	1.45	1.60	0.30	0.70	1.00	1.40	1.60	1.70
4	0.40	0.80	1.20	1.50	1.70	1.90	0.40	0.80	1.20	1.60	1.80	2.10
6	0.60	1.20	1.70	2.00	2.25	2.45	0.60	1.30	1.80	2.20	2.41	2.60
8	0.90	1.60	2.15	2.50	2.75	2.90	0.95	1.60	2.30	2.60	3.00	3.20
10	1.20	2.00	2.60	2.95	3.19	3.40	1.20	2.00	2.50	2.80	3.40	2.70
12	1.40	2.35	3.00	3.40	3.60	3.80	1.40	2.41	2.90	3.60	3.80	4.09
14	1.70	2.70	3.40	3.75	3.94	4.17	1.80	2.80	3.49	3.98	4.29	4.61
16	1.90	3.10	3.75	4.09	4.29	4.61	2.00	3.19	3.98	4.61	4.69	5.00
20	2.35	3.65	4.29	4.69	4.88	5.20	2.40	3.90	4.61	5.12	5.39	5.59
24	2.80	4.09	4.69	5.20	5.39	5.67	2.80	4.37	5.08	5.59	5.91	6.10
28	3.00	4.45	5.20	5.59	5.79	5.98	3.10	4.80	5.47	6.10	6.34	6.54
32	3.19	4.72	5.51	5.91	6.10	6.34	3.40	5.04	5.79	6.38	6.69	6.89
	r/R = 0.5						r/R = 0.7					
JD	H=0.2	0.5	1.0	2.0	5.0	∞	H=0.2	0.5	1.0	2.0	5.0	∞
1	0.10	0.20	0.45	0.70	0.90	1.20	0.10	0.20	0.40	0.80	1.20	1.70
2	0.20	0.50	0.70	1.00	1.30	1.60	0.20	0.50	0.90	1.30	1.80	2.41
3	0.30	0.70	1.00	1.40	1.70	2.00	0.30	0.70	1.20	1.70	2.20	2.81
4	0.40	0.90	1.30	1.65	2.00	2.30	0.40	0.90	1.50	2.15	2.75	3.30
6	0.65	1.30	1.70	2.41	2.76	3.00	0.65	1.45	2.20	2.41	3.70	4.17
8	0.90	1.70	2.35	2.90	3.19	3.60	0.90	2.00	2.95	3.80	4.49	5.00
10	1.20	2.00	2.90	3.49	3.80	4.17	1.20	2.60	3.60	4.53	5.28	5.90
12	1.50	2.65	3.40	3.98	4.29	4.69	1.55	3.00	4.29	5.28	6.10	6.57
14	1.80	3.10	3.90	4.49	4.80	5.20	1.80	3.60	5.00	5.98	6.77	7.28
16	2.10	3.40	4.29	4.88	5.24	5.59	2.06	4.09	5.59	6.59	7.40	7.99
20	2.55	4.09	5.12	5.59	5.98	6.30	2.60	4.88	6.97	7.68	8.39	9.06
24	3.00	4.65	5.7	5.83	6.54	6.89	3.10	5.59	7.40	7.40	9.29	9.88
28	3.30	5.12	6.10	6.06	7.01	7.36	3.55	6.14	7.99	9.21	9.88	--
32	3.60	5.39	6.38	6.30	7.40	7.72	3.90	6.69	8.58	9.69	--	--
	r/R = 0.9						r/R = 1.0 (surface)					
JD	H=0.2	0.35	0.5	0.7	1.0	2.0	H=0.2	0.35	0.5	0.7	1.0	1.5
1	0.10	0.20	0.25	0.40	0.50	1.30	0.10	0.20	0.30	0.40	0.60	1.40
2	0.20	0.40	0.50	0.60	1.00	2.60	0.20	0.35	0.50	1.00	2.00	--
3	0.30	0.60	0.80	1.30	1.75	3.60	0.30	0.50	0.80	1.70	3.98	--
4	0.40	0.76	1.00	1.60	2.35	4.69	0.40	0.77	1.20	2.50	--	--
6	0.60	1.20	1.80	3.00	4.37	7.01	0.77	1.30	2.50	--	--	--
8	0.91	1.70	2.60	4.61	6.10	8.58	1.00	1.90	5.00	--	--	--
10	1.25	2.30	3.49	6.38	7.99	--	1.40	3.00	--	--	--	--
12	1.60	3.00	4.69	7.99	9.50	--	1.75	4.61	--	--	--	--
14	2.00	3.80	5.91	9.60	--	--	2.20	10.00	--	--	--	--
16	2.40	4.69	7.28	--	--	--	2.50	--	--	--	--	--
20	3.10	6.14	10.00	--	--	--	3.49	--	--	--	--	--
24	3.70	7.40	--	--	--	--	4.37	--	--	--	--	--
28	4.17	8.58	--	--	--	--	5.91	--	--	--	--	--
32	4.69	9.50	--	--	--	--	6.38	--	--	--	--	--

Table 6-7. Approximate Severity of Quench, H.

Agitation	Air	Oil	Water	Brine
None	0.02	0.3	1.0	2.0
Moderate		0.5	1.5	3.5
Violent		1.0	4.0	7.0
Spray		1.3	7.0	9.0

Table 6-8. Effect of Carbon Content, Grain Size, and Alloy on the Hardenability of Steel

A. Multiplying factors for carbon (%) and grain size

1. $f_c = 0.41\ C + 0.15$ GS = #2
2. $f_c = 0.35\ C + 0.132$ GS = #4
3. $f_c = 0.32\ C + 0.123$ GS = #5
4. $f_c = 0.29\ C + 0.116$ GS = #6
5. $f_c = 0.27\ C + 0.107$ GS = #7
6. $f_c = 0.25\ C + 0.098$ GS = #8
7. $f_c = 0.21\ C + 0.080$ GS = #10

B. Multiplying factors for alloying element (%)

1. $f_B = 17.23\ B^{0.268}$ B < .002
2. $f_{Mn} = 3.46\ Mn + 1$ 0 < Mn < 1.2
3. $f_{Mn} = 5.125\ Mn - 1$ 1.2 < Mn < 2.0
4. $f_{Mo} = 3.09\ Mo + 1$ Mo < 1.0
5. $f_{Cr} = 2.18\ Cr + 1$ Cr < 2.0
6. $f_{Si} = 0.7\ Si + 1$ Si < 2.0
7. $f_{Ni} = 0.4\ Ni + 1$ Ni < 2.0

Table 6-8B lists the alloying elements factors that must be multiplied by the carbon content factor to obtain the ideal critical diameter of the alloy steel. If two or more alloying elements are used, the product of their individual factors must be multiplied by the carbon factor. For 1% manganese the factor f_{Mn} is 4.46; for 1/4% silicon the factor f_{Si} is 1.18.

164   Chapter 6

Thus, if a 0.40% carbon steel having a grain size #2 contains 1% manganese and 1/4% silicon, the ideal critical diameter is $DI = 0.31 \times 4.46 \times 1.18 = 1.65$ in.

The multiplying factors in Table 6-8B are listed in order of decreasing benefit. Boron is the most beneficial. It has an exponential effect up to a maximum of 0.002% boron. All of the other elements have a linear effect. Manganese is the next most beneficial element; it is also one of the least expensive. Of the common alloying elements, nickel is the least beneficial as far as impacting hardenability.

Table 6-9 lists the dividing factors for estimating the hardness of JDs from 1/4" (4 JD) to 2" (32 JD) as a function of the ideal critical diameter of any steel. For example, consider a steel having 0.40% carbon and a DI of 3.0." It has a primary martensite hardness of 73 + 17 ln 0.4 or 57 R_C. The dividing factor fd is 1.07 at 4 JD (Table 6-8). Therefore the hardness at this JD is 57/1.07 or 53 R_C. And at 8 JD it is 57/1.36, or 42 R_C.

The hardenability band of any steel can thus be estimated from its composition and grain size. To obtain the upper limit of the hardenability band, one uses the maximum values of the range of grain size and alloys. To obtain the lower limit, the minimum values are used.

The limits of the hardenability band computed in this manner are outside the experimentally determined values since in practice not all of the alloying elements nor the grain size are at their maximum or minimum values.

An example of the use of this method, the minimum of the hardenability band of AISI 4161 steel is calculated. This steel is not an AISI standard and is not listed in Table 6-5. Assume that the austenite grain size may vary from 2 to 8. The AISI range of composition for 4161 steel is .56/.64 C, .75/1.00 Mn, .70/.90 Cr, .25/.35 Mo, and .15/.25 Si. The calculations are made as follows:

1. For primary martensite: $R_C = 73 + 17 \ln 0.56 = 63$
2. Multiplying factor for carbon and grain size:
   $f_c = 0.25 \times 0.56 + 0.098 = 0.24$
3. Multiplying factors for the alloying elements:
   $f_{Mn} = 3.46 \times 0.75 + 1 = 3.60$
   $f_{Mo} = 3.09 \times 0.25 + 1 = 1.77$
   $f_{Cr} = 2.18 \times 0.70 + 1 = 2.53$
   $f_{Si} = 0.7 \times 0.15 + 1 = 1.11$

Thermal Strengthening and Costs   165

Table 6-9. Dividing Factor for Ideal Critical Diameters at Various Jominy Distances (JD)

DI	JD= 4	8	12	16	20	24	28	32
1.0	2.4	3.33						
1.5	1.48	2.35	3.0	3.3	3.5	3.7	3.9	4.1
1.6	1.41	2.19	2.8	3.14	3.35	3.51	3.69	3.88
1.7	1.35	2.08	2.64	2.99	3.2	3.35	3.51	3.68
1.8	1.3	1.95	2.49	2.82	3.06	3.2	3.35	3.48
19	1.26	1.86	2.33	2.7	2.92	3.05	3.2	3.3
2.0	1.23	1.77	2.24	2.6	2.81	2.93	3.06	3.16
2.1	1.2	1.71	2.17	2.5	2.71	2.81	2.94	3.04
2.2	1.18	1.65	2.09	2.4	2.61	2.72	2.84	2.93
2.3	1.16	1.6	2.01	2.32	2.52	2.64	2.75	2.84
2.4	1.14	1.55	1.95	2.26	2.47	2.57	2.69	2.76
2.5	1.13	1.52	1.9	2.19	2.39	2.51	2.61	2.69
2.6	1.11	1.49	1.84	2.13	2.32	2.44	2.55	2.62
2.7	1.1	1.45	1.8	2.09	2.27	2.39	2.49	2.56
2.8	1.09	1.42	1.75	2.02	2.22	2.34	2.42	2.5
2.9	1.08	1.39	1.7	1.99	2.16	2.28	2.36	2.44
3.0	1.07	1.36	1.66	1.93	2.11	2.22	2.31	2.38
3.1	1.06	1.33	1.62	1.89	2.06	2.18	2.27	2.33
3.2	1.05	1.3	1.6	1.85	2.01	2.13	2.22	2.29
3.3	1.04	1.28	1.56	1.8	1.97	2.09	2.17	2.23
3.4	1.03	1.26	1.53	1.76	1.93	2.04	2.13	2.19
3.5	1.025	1.25	1.51	1.73	1.9	2.0	2.09	2.14
3.6	1.02	1.235	1.49	1.7	1.86	1.97	2.05	2.1
3.7	1.015	1.22	1.46	1.67	1.82	1.93	2.01	2.06
3.8	1.01	1.205	1.44	1.63	1.78	1.89	1.97	2.01
3.9	1.005	1.19	1.41	1.6	1.75	1.84	1.92	1.97
4.0	1.005	1.175	1.39	1.56	1.72	1.8	1.88	1.93
4.1	1.0	1.16	1.36	1.53	1.69	1.77	1.84	1.9
4.2	1.0	1.145	1.33	1.5	1.65	1.73	1.8	1.86
4.3	1.0	1.13	1.305	1.47	1.61	1.7	1.77	1.82
4.4	1.0	1.12	1.29	1.45	1.58	1.66	1.73	1.78
4.5	1.0	1.11	1.27	1.42	1.54	1.62	1.7	1.75
4.6	1.0	1.105	1.25	1.4	1.51	1.59	1.66	1.71
4.7	1.0	1.095	1.23	1.38	1.49	1.55	1.62	1.68
4.8	1.0	1.09	1.21	1.35	1.46	1.52	1.595	1.63
4.9	1.0	1.08	1.2	1.325	1.42	1.5	1.55	1.6
5.0	1.0	1.075	1.185	1.3	1.4	1.47	1.51	1.56
5.2	1.0	1.065	1.15	1.26	1.34	1.4	1.45	1.49
5.4	1.0	1.05	1.11	1.21	1.29	1.34	1.39	1.42
5.6	1.0	1.04	1.09	1.17	1.24	1.29	1.32	1.37
5.8	1.0	1.03	1.06	1.125	1.19	1.23	1.275	1.3
6.0	1.0	1.02	1.025	1.1	1.15	1.195	1.22	1.255
6.2	1.0	1.01	1.01	1.07	1.11	1.14	1.16	1.2
6.4	1.0	1.0	1.0	1.04	1.09	1.105	1.12	1.155
6.6	1.0	1.0	1.0	1.01	1.05	1.05	1.08	1.12
6.8	1.0	1.0	1.0	1.0	1.03	1.03	1.04	1.07
7.0	1.0	1.0	1.0	1.0	1.01	1.01	1.01	1.03

4. Ideal critical diameter:
   $DI = 0.24 \times 3.60 \times 1.77 \times 2.53 \times 1.11 = 4.29$
5. The dividing factors (Table (6-9)) and resulting hardness:

JD =	0.0	4.0	8.0	12.0	16.0	20.0	24.0	28.0	32.0
fd =	1.0	1.0	1.13	1.31	1.47	1.61	1.70	1.77	1.82
R_C =	63	63	56	48	43	39	37	36	35

As stated previously this calculated lower limit of the hardenability band is lower than the one that would be experimentally determined.

COST OF STEEL

In order to select the optimal steel on the basis of strength, it is necessary to have the cost information for the various steels as well as the hardenability data. Table 6-10 summarizes the costs for plain carbon steel. It should be noted that there are two quality levels of plain carbon steel. The top portion of the table is for special, or the standard quality, which has a base price of 30.20¢/lb. The bottom portion of the table is for Merchant Quality, which is a lower quality steel and has a base price of 28.50¢/lb.

To illustrate the use of the cost data in Table 6-10, the cost of AISI 1040 special quality 2" diameter hot-rolled rounds with random lengths in lots of 20,000 lbs. The base price is 30.20¢/lb. The size extra is 0.75¢/lb. The quantity extra is 0.70¢/lb. No quality extra is specified. The cost extra for 1040 steel is 1.05¢/lb. Therefore the cost is 32.70¢/lb, F.O.B. Johnstown, PA. That is, the price does not include the cost of the freight or transportation from the mill in Johnstown to the plant where it is to be used.

Table 6-11 summarizes the cost of a variety of hot-rolled alloy bars calculated as base price, plus the extra cost of the alloy for some of the more common standard alloy grades. To compare the cost of alloy steel to plain carbon, consider the order in the preceding example but with 4140 steel substituted for 1040 steel. For the alloy steel the base price is 33.50¢/lb. Thus the price of 4140 hot rolled steel is 37.90¢/lb, compared with 32.70¢/lb for the 1040 steel. The price of AISI 4340 steel is 43.50¢/lb, which is 33% greater than the price of 1040 steel.

As may be expected, the cheapest of the standard alloy steels are the 13zz steels since they contain manganese as the only alloying element.

Table 6-10. Cost of Carbon Steel (cents per pound)

Hot-Rolled Carbon Bars: Special Quality
Base = 30.20 (Johnstown, PA) Random lengths

Size extra for rounds. (dia. in inches)

size:	3/8	1/2	1	2	4	8
extra:	6.00	2.70	0.90	0.75	0.80	1.50

Quantity Extra: over 40,000 lbs = base; 20,000 to 39,999 lbs = 0.35 extra; 10,000 to 19,999 lbs = 0.70 extra

Quality Extra: For special internal soundness, special surface, non-metallic inclusions (microscopic) or special hardenability = 0.35 extra.

Extra for Standard Alloy

AISI	extra	AISI	extra	AISI	extra	AISI	extra
1005	0.80	1060	1.05	1547	1.55	1117	2.45
1010	0.40	1070	0.95	1551	1.25	1141	1.80
1015	1.00	1080	0.95	1566	1.15	1144	2.05
1020	1.00	1090	0.95	1038H	1.05	1212	1.50
1030	1.05	1513	1.35	1045H	1.05	1213	2.05
1040	1.05	1522	1.85	1541H	1.55	1213B	2.05
1045	1.05	1536	1.55	15B37H	2.70	12L14	2.90

Hot Rolled Carbon Bars: Merchant Quality
Base = 28.50 (Johnstown, PA). Random lengths

Size extra for rounds. (dia. in inches)

size:	1/2	9/16	5/8	3/4	1	2
extra:	2.70	2.20	2.00	1.35	0.90	1.25

Quantity Extra. over 40,000 lbs = base;
20,000 to 40,000 lbs = 0.35 extra; 10,000 to 20,000 lbs = 0.70 extra

Chemical Requirement Extra: Basic Open Hearth (BOH)
Includes: M1008, M1010, M1012, M1015, M1017, M1020, M1023, M1025, M1031, M1044 (all are extra)

(Approximate 1996 mill prices. Data compiled from 1977 Bethlehem Steel Catalog and adjusted for inflation)

Table 6-11. Cost of Alloy Steel (cents per pound)

Base = 33.50* (Johnstown, Pa.)
Size extra for rounds. (dia. in inches)

size:	1/2	1	2	4	8
extra:	2.70	0.90	0.75	0.80	1.50

Quantity extra: over 40,000 lbs = base; 20,000 to 39,999 lbs = 0.35 extra; 10,000 to 19,999 lbs = 0.70 extra

Extra for Standard Alloys

AISI	Extra	AISI	Extra	AISI	Extra	AISI	Extra
1330	1.25	4161	3.15	5130	1.90	8640	4.15
1345	1.25	4320	8.70	5140	1.55	8655	4.15
4012	3.10	4340	8.55	5150	1.55	8740	4.40
4023	2.85	4620	8.30	5160	1.55	9255	3.00
4047	2.00	4720	6.80	6150	4.10	9260	3.00
4130	2.70	4820	14.10	8615	5.80	50B44	2.40
4140	2.95	5015	2.45	8620	4.95	50B60	2.40
4150	2.95	5120	2.50	8630	4.15	51B60	2.65

*For BOH (basic open earth) random lengths. Electric furnace steel is 1.25 to 2.00 extra

Extra for Nonstandard Alloys: Manganese (with no Chromium)

max. of Mn, range (%)	Carbon Content				
	0-0.11	0.11-0.20	0.21-0.24	0.24-0.28	>0.29
0.41-0.70	0.70	0.60	0.45	0.35	0.35
0.71-1.00	1.30	0.95	0.50	0.45	0.45
1.01-1.30	1.75	1.30	0.60	0.60	0.60
1.31-1.60	2.15	1.70	1.00	0.75	0.75
1.61-1.90	2.70	2.20	1.40	1.05	1.00
1.91-2.20	3.30	2.70	1.80	1.45	1.25

Silicon: up to 0.35%Si = Base

For Silicon between 0.35% and 2.75%: $CSi = 0.78\, Si^{1.36}$ where CSi = cost extra and Si = % of silicon

Nickel (0 to 10%): $CNi = 3.25\, Ni^{1.04}$

Molybdenum (0 to 3.9%): $CMo = 6.0\, Mo^{1.0}$

Boron (up to 0.002%): $CB = 1.10$

(Approximate 1996 mill prices. Data compiled from 1977 Bethlehem Steel Catalog and adjusted for inflation)

Thermal Strengthening and Costs 169

The bottom portion of Table 6-11 summarizes the cost of the individual alloying elements when non-standard grades are specified. The base price is the same, 33.50¢/lb. To illustrate the relative costs of the common alloying elements consider the cost extra for 1% of each element for a 0.40% carbon steel. The extra for manganese is 0.45¢/lb. For silicon it is 0.78¢/lb. Nickel and molybdenum are 3.25 and 6.00¢/lb, respectively. However, molybdenum is seldom added in excess of 1/2% whereas the nickel content in some standard steels is as high as 4%. The price extra for boron is 1.10¢/lb for any amount up to the maximum amount that is used, 0.002%.

The prices in Tables 6-10 to 6-14 are approximate values, and only a few of the alloys are listed. The values included here are to be used for preliminary design only. The material suppliers should be contacted for firm prices.

NON-FERROUS AND NON-METALLIC MATERIALS

At the present time there are very few quantitative relationships between the mechanical properties and the microstructures of all the non-ferrous materials. As a consequence, it is still often necessary to determine the properties of these materials from unrelated tabular data. For each specific material, the properties are listed for the common conditions that the material is used such as annealed, cold drawn, or aged at some specific temperature. The basic mechanical properties of a large number of non-ferrous materials, including plastics, are given in Appendix B. The reader should contact the material producer for information concerning the properties of materials not included there.

Table 6-12 gives cost information for round bars of several grades of aluminum alloy. By utilizing both the strength data given in Appendix D and the cost data given in this table, one can compare the cost of a round part made of aluminum alloy to the cost of a similar part made of steel on the basis of equal load-carrying capacity.

Table 6-13 gives similar data for some of the common copper base alloys, and Table 6-14 gives the costs of some of the common plastic materials. In Table 6-14 the cost is expressed as dollars per foot for all the materials except for the extruded acrylic rods in which case the cost is given in dollars per pound.

Table 6-12. Cost of Aluminum Bars (Dollars per lb)

Non-heat treatable plates (40 ton lots)

thickness (in.)	1100-H14	3003-H14	5052-H34
1/4	2.08	2.10	2.14
1	2.02	2.07	2.12

Heat treatable plates (40 ton lots)

thickness (in.)	2014-T4	2024-T4	6061-T4	7075-T6
1/4	3.30	3.37	2.46	3.49
1	3.08	3.10	2.38	3.21

Cold finished bars (40 ton lots)

thickness (in.)	2011-T3	2014-T4	2024-T4	6061-T4	7075-T6
1/2	2.36	3.42	3.36	2.72	3.63
1	2.24	3.34	3.29	2.66	3.55
2	2.24	3.22	3.18	2.46	3.47
3	2.30	3.25	3.25	2.52	3.50
6	2.48	3.29	3.35	2.67	3.55

(Data compiled from 1996 Kaiser Aluminum catalog.)

Table 6-13. Cost of Some Copper Alloys (Dollars per lb)

Alloy	Composition	Size	Price
Cu 110	99.90% min. Cu	1" round	2.37
Cu 110	99.90% min. Cu	1" plate	3.18
Cu 102	99.95% min. Cu (O_2 free)	1" round	7.40
Br 220	90 Cu, 10 Zn	1" round	9.12
Br 260	70 Cu, 10 Zn	1/8" plate (soft)	5.50
Br 260	70 Cu, 30 Zn	1/8" plate (3/4 hard)	5.50
Br 360*	61.5 Cu, 28.5 Zn, 3 Pb	1" round	6.59
Bz 510	95 Cu, 5 Sn	1" round	9.89

*Free cutting brass
1996 prices from various sources (2,000 lb. lots).

Table 6-14. Cost of Some Common Plastics

Plexiglass (acrylic)
    1/4 in. thick, 4 ft. × 8 ft. sheet $70.93/sheet
    1" dia. extruded rod $1.59/ft.
    1" dia. cast rod $3.64/ft.

ABS (acrylonitrile butadiene styrene)
    1/4 in. thick, 4 ft. × 8 ft. sheet
    black, smooth on both sides $73.96/sheet

PVC (polyvinyl chloride) Type 1, gray
    1/4 in. thick, 4 ft. × 8 ft. sheet $63.89/sheet
    1 in. dia. rod $1.39/ft.

Nylon (polymide) white
    1/4 in. thick, 4 ft. × 8 ft. sheet $102.14/sheet
    1 in. dia. rod $2.67/ft.

Teflon (polytetrafluoroethylene) white
    1/4 in. thick, 4 ft. × 8 ft. sheet $231.66/sheet
    1 in. dia. rod $8.97/ft.

Data compiled from 1996 Cadillac plastic catalog (prices pertain to large quantity purchases).

EXAMPLE PROBLEMS

6-1. A cylindrical bar 24 in. long is needed that will support a load of 50,000 lb with a factor of safety of 2 based on the yield strength. If 9262 H steel costs $0.65/lb and 4140 H steel costs $0.60/lb, which would be the more economical to use? Use the minimum values of the hardenability limits in Table 6-5. Assume oil quench with moderate agitation.

172   Chapter 6

Solution:
Use the relationships in Table 6-2. For 9262 H steel and up to 4 JD the Rockwell C is 60. $60R_c = 653H_B$. Stress relieving the steel at 400°F gives

$$H_B = 0.95 \times 653 = 620$$
$$S_u = 0.5 \times 620 = 310 \text{ ksi}$$
$$S_y = 1.05 \times 310 - 30 = 295 \text{ ksi}$$

For a factor of safety of 2, use a load of 100,000 lb. Therefore the area required is $100 \text{ klb} / 295 \text{ ksi} = 0.339 \text{ in.}^2$, or a diameter of 0.657."

From Table 6-6 it can be seen that for H = 0.5, the center of a 0.657" diameter bar cools at rate equal to 3.5 JD. Therefore, the 0.657" diameter bar will have a yield strength of 295 ksi.

In similar fashion, for the 4140 H, since the carbon content is lower, the strength will also be lower than for the 9262 H steel. Therefore, try a Rockwell C of 47 at 8 JD. The H_B is 444. The stress-relieved H_B will be approximately 0.96 × 444 = 426. (0.96 is used instead of 0.95 because the microstructure is a little less than 100% martensite.) This gives $S_u = 213$ and $S_y = 194$. Area required is $100/194 = 0.515 \text{ in.}^2$, or 0.810" diameter. From Table 6-6 it is evident that the center cools at a rate equal to 4 JD.

From Table 6-5 the minimum hardness at 4 JD is 51 R_C or 495 H_B. Stress relieving will reduce the H_B to 0.95 × 495 = 470. $S_u = 235$ and $S_y = 217$.

The required area = $100/217 = 0.461 \text{ in.}^2$, or 0.766 in. diameter. The cost for a 24" long bar (based on large quantities) is:
for 9262 H: $\$0.65 / \text{lb} \times 0.283 \text{lb} / \text{in.}^3 \times 24 \times 0.339 \text{in.}^3 = \$1.50$
for 4140 H: $\$0.60 / \text{lb} \times 0.283 \text{lb} / \text{in.}^3 \times 24 \times 0.461 \text{in.}^3 = \$1.88$

6-2. A certain heat of steel has the following composition: 0.40%C, 1.0%Mn, 0.25%Si, 1%Ni, and a grain size of #5.

Thermal Strengthening and Costs

a. Calculate the Jominy end-quench data for this steel.
b. Calculate the amount of microconstituents that would be present at the center if a 2" diameter bar of this steel were cooled from the austenite region with a severity of quench equal to 1.0.

Solution:
a. $fc = 0.32 \times 0.4 + 0.123 = 0.251$ $\quad f_{Mn} = 3.46 \times 1 + 1 = 4.46$
$f_{Si} = 0.7 \times 0.25 + 1 = 1.175$ $\quad f_{Ni} = 0.4 \times 1 + 1 = 1.400$
$DI = 0.251 \times 4.46 \times 1.175 \times 1.4 = 1.842"$
Jominy Curve $\quad (R_C) 0.4C = 73 + 17 \ln 0.4 = 57.4$

JD	0	4	8	12	16	20	24	28	32
f_D	1.0	1.28	1.90	2.42	2.76	2.94	3.13	3.28	3.40
R_C	57.4	44.8	30.2	23.7	20.8	19.5			

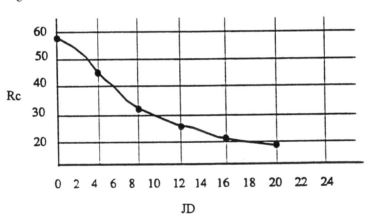

b. Center of 2" Dia,
$H = 1.0$ gives $JD = 7.5$, $R_C = 31$, and $H_B = 293$
$C_E = 0.8 - 0.061 \times 1 \, Mn - 0.125 \times 0.25 \, Si - 0.03 \times 1 \, Ni = 0.68$
$f_P = 0.4 / 0.68 = 0.59$ $\quad f_\alpha = 0.41$
If all $\alpha + P_f$ $\quad H_B = 0.59 \times 380 + 0.41 \times 80 = 257$
There must be a small amount of martensite of 0.68%C
Let X = fraction of M and $f_\alpha = 0.41$

then $f_P + f_M = 0.59$ and $f_P = 0.59 - X$

$$293 = f_\alpha \times (H_B)_\alpha + (f_M)(H_B)_M + (f_P)(H_B)_P$$
$$= 0.41 \times 80 + 741X + 380(0.59 - X)$$
$$= 32.8 + 741X + 224.2 - 380X$$
$$36 = 361X \quad X = 0.10, \text{ or } 10\%\,M$$

The composition is: 41% α and 49% P_f and 10%M

REFERENCES

1. Van Vlack, L.H. *Elements of Materials Science*, 2nd ed. Addison Wesley, Reading, MA, 1989.
2. Flin, R.A. and P.K. Trojan, *Engineering Materials and Their Application,* Houghton Mifflin, Boston, 1990.
3. United States Steel, *Atlas of Isothermal Transformation Diagrams*, United States Steel Co., Pennsylvania, 1951.
4. Datsko, J. *Material Properties and Manufacturing Processes.* John Wiley & Sons, New York, 1967.
5. Hollomon, J.H. and L.D. Jaffe, *Ferrous Metallurgical Design.* John Wiley & Sons, New York, 1947.
6. Craft, W., and J.L. Lamont, *Hardenability and Steel Selection.* Pitman Publishing, New York, 1949.
7. Bethlehem Steel Co. *Modern Steels and their Properties,* Bethlehem Steel Co., Bethlehem, PA, 1972.
8. Grossman, M.A., *Principles of Heat Treatment,* Haddon Craftsmen, Pennsylvania, 1935.

Thermal Strengthening and Costs

STUDY PROBLEMS

6-1. A 4" dia. × 18" long cylinder of 6150 H steel is quenched in water with moderate agitation. Based on the minimum values of hardenability:
  a. Estimate the hardness at the surface.
  b. Estimate the tensile and yield strengths at the center.
  c. What diameter bar, when oil quenched with moderate agitation, would have the same hardness at the center? Explain.

6-2. What minimum axial load would cause yielding in a 2 in. diameter bar of 1340 H steel oil-quenched with moderate agitation and tempered at 400°F?

6-3. What is the critical cooling rate for an E4340 H steel? For 1080 steel?

6-4. Four small pieces of 1050 steel are heated to 1340°F and cooled as indicated below. Determine the amount of microconstituents and give the hardness for each process:
  a. Furnace cool
  b. Air cool
  c. Oil quench
  d. Water quench

6-5. Select a steel to carry a uniaxial load of 10^6 lbs on the basis of maximum S_y. Minimum cost is desired, and a factor of safety of 1.

6-6. Select a steel to transmit 5 hp @ 1 rpm on the basis of maximum S_y. Minimum cost is desired, and a factor of safety of 1.

6-7. To fabricate a certain structural component, the following processing steps were performed:
  a. 1/8" thick sheets of 2024-T4 aluminum were purchased.
  b. Preform blanks were cut (sheared) from the sheets.
  c. Prior to forming in "stamping" dies on a punch press the blanks were solution heat treated (W condition) and kept in cold water until the parts were formed.
  d. After forming, the parts were allowed to age naturally.
What is the proper designation of the aluminum in these finished parts?

Chapter 7

MECHANICAL PROPERTIES

INTRODUCTION

Having gained an understanding of atomic structures and microstructures, the reader is now in a position to interpret and critically evaluate data on the mechanical properties of metals and other materials. The ability to evaluate whether published data is reliable and valid and whether the data for a given material is the best that can be obtained is essential for optimal materials selection. This chapter focuses on the interrelationships of the various mechanical properties.

Most mechanical properties are structure sensitive, that is, they are affected by changes in either the lattice structure or the microstructure, but the modulus of elasticity is one property that is structure insensitive. For example, the strength, ductility, and toughness of any material (regardless of whether it is a pure element such as copper, a simple alloy such as AISI 1080 steel, or a complex one such as cobalt-base superalloy) varies with the grain size, amount of cold work if any, or the microstructure if heat treated. The modulus of elasticity of any material (17×10^3 ksi for copper, 30×10^3 ksi for 1080 steel, 32.6×10^3 ksi for a 50% cobalt-base superalloy) is the same, regardless of the grain size, amount of cold work, or microstructure.

The mechanical properties are discussed individually in the following sections. Many new quantitative relationships for the properties are presented here. These quantitative relationships make it possible to understand the mechanical properties to a depth that is not possible by simply comparing materials data in conventional tabular form.

HARDNESS

Hardness is used by the design engineer more frequently than any other of the mechanical properties to specify the final condition of a structural part. This is due in part to the fact that the hardness tests are inexpensive and not time consuming to conduct. The tests can be performed on a finished part without the need for machining a special test specimen. In other words, a hardness test may be a nondestructive test in that it can be performed on the actual part without affecting its service function.

Hardness is frequently defined as a measure of a material's ability to resist plastic deformation or penetration by an indenter having a spherical or conical end. In the sense that hardness tests are practical shop tests rather than basic scientific tests, hardness is more of a technological property of a material than it is a scientific or engineering property. All of the hardness scales in use today, including those such as the Brinell that have units of stress (kg/mm^2) associated with them, give relative values rather than absolute ones. Thus a hardness of 50 R_C has meaning only in comparison to other values on the same scale—it is harder than a material of 25 R_C but not twice as hard. How hard must a material be in order to have a hardness that is twice that of a material having a hardness of zero Rockwell B? In the sense of being relative, these hardness scales are similar to the Fahrenheit and Celsius temperature scales. There are absolute temperature scales, namely the Rankine and Kelvin scales, but there are no absolute hardness scales at the present time. Even though some hardness scales, such as the Brinell, have units of stress associated with them, they are not absolute scales because a given piece of material (such as a 2" cube of brass) will have different Brinell hardness numbers depending upon whether a 500 kg or a 3000 kg load is applied to the indenter. Table 7-3 (page 218) is a conversion chart for the common hardness scales.

Rockwell Hardness

Rockwell hardness is determined by an indentation test that measures the depth of indentation from an increment of load. The Rockwell scales are by far the most frequently used hardness scales in industry even though they are relative.

The reasons for their widespread acceptance are the simplicity of the testing apparatus, the short time necessary to obtain a reading, and the ease with which reproducible readings can be obtained. The last reason is due in part to the fact that the testing machine has a "direct reading" dial, that is, a needle points directly to the actual hardness value without the need of referring to a conversion table as must be done with the Brinell, Vickers, and Knoop hardness tests. The chart below lists the most common Rockwell hardness scales.

Scale	A	B	C	D	E	F	G	H	K
Indenter	1	2	1	1	3	2	2	3	3
Load (kg)	60	100	150	100	100	60	150	60	150

Indenter #1 is a diamond cone having an included angle of 120° and a spherical end radius of 0.008." Indenters #2 and #3 are 1/16" and 1/8" diameter steel balls respectively. In addition to the above scales there are several others for testing very soft bearing materials, such as babbitt, that use 1/4" and 1/2" diameter balls. Also there are several "superficial" scales that use a special diamond cone with loads less than 50 kg to test the hardness of surface-hardened layers.

Each scale is used for different types of materials: the A scale for extremely hard materials such as carbides or thin case-hardened layers on steel; the B scale for soft steels, copper and aluminum alloys, and soft cast irons; the C scale for medium and hard steels, hard cast irons, and hard nonferrous alloys; and the E and F scales for soft copper and aluminum alloys. The remaining scales are used for even softer alloys and for non-metals as well.

Several precautions must be observed in the proper use of the Rockwell scales. The ball indenter should not be used for any material having a hardness greater than about 50 R_C, otherwise the steel ball will be plastically deformed or flattened and thus give erroneous readings. Readings taken on the sides of cylinders, or spheres, should be corrected for the curvature of the surface. Readings of less than 20 on the C scale should not be recorded or specified because they are unreliable and subject to much variation because the depth of the indentation corresponding to a value of 20 R_C is 0.008," which is the same as the radius on the indenter. Indentations are spherical for hardnesses greater than 20 R_C, but they are not completely spherical for lower hardnesses.

Rockwell hardness tests are conducted by applying a spring load (called the minor load) of 10 kg to the indenter and then zeroing the direct reading dial. Then the appropriate additional dead load (called the major load) is applied. The major load for each scale is 10 kg less than the total load listed in the preceding chart. For example, the major load is 140 kg for the C scale. About

10-15 seconds after the needle comes to rest, the major load is retracted while the minor load is still applied. The position of the needle on the dial gauge is recorded as the hardness of the material on the appropriate scale. Finally, the minor load is released and the part withdrawn from the testing machine. Thus, the Rockwell hardness number is the difference in the penetration of the indenter when the total load and the minor load are applied.

Actually, the hardness numbers for all the Rockwell scales are an inverse measure of the depth of the indentation. Each division on the Rockwell machine's dial gage corresponds to an 8×10^{-7} inch depth of penetration. The penetration with the C scale varies between 0.0005" for hard steel and 0.0015" for very soft steel when only the minor load is applied. The total depth of penetration with both the major and minor loads applied varies from 0.003" for the hardest steel to 0.008" for soft steel (20 R_C). Since these indentations are relatively shallow, the Rockwell C hardness test is considered a nondestructive test and it can be used on fairly thin parts.

Although negative hardness readings can be obtained on the Rockwell scales, they are usually not recorded as such, but rather a different scale is used that gives readings greater than zero. The only exception to this is when one wants to show a continuous trend in the change in hardness of a material due to some treatment. A good example of this is the effect of cold work on the hardness of a fully annealed brass. Here the annealed hardness may be $-20\ R_B$ (Rockwell B) and increase to 95 R_B with severe cold work.

Brinell Hardness

Brinell hardness is determined by dividing the load that is applied to a spherical indenter by the surface area of the spherical indentation produced and has the units of kg/mm^2. In this book it is identified by the symbol H_B. Most readings are taken with a 10 mm ball of either hardened steel or tungsten carbide. The loads that are applied vary from 500 kg for soft materials to 3000 kg for hard materials. The steel ball should not be used on materials having a hardness greater than about 525 H_B (52 R_C) because of the possibility of deforming the ball and making it inaccurate for further use.

The Brinell hardness machine is as simple, though more massive, than the Rockwell hardness machine, but the standard model is not direct reading and requires a longer time to obtain a reading. In addition, the indentation is much larger than the Rockwell's — it is considered a destruction test — and it cannot be used on hard steel. The method of operation is simple. The prescribed load is applied to the 10 mm diameter ball for approximately 20 seconds. The part is then withdrawn from the machine and the operator measures the diameter of the indentation by means of a millimeter scale etched on the eyepiece of a Brinell microscope. The Brinell hardness number is then obtained from the equation

$$H_B = \frac{L}{\frac{\pi D}{2}(D - \sqrt{D^2 - d^2})} \qquad (7\text{-}1)$$

where L is the load in kilograms, D is the diameter of the indenter, and d is the diameter of the indentation. (The denominator is the spherical area of the indentation.) The operator usually does not have to calculate the hardness number from this equation because each machine is provided with a chart that converts the diameter of indentation to the appropriate hardness number. A portion of such a chart for 500 kg and 3000 kg load is included in Table 7-3 (pg. 218). Some newer models are available with direct digital readout.

Brinell developed his hardness testing machine in 1900 more as a technological test than a scientific one. He believed that industry needed a simple test that would give a single reproducible hardness number. The Brinell hardness test has proved to be very successful, due partly to the fact that for some materials the results could be directly correlated with tensile strength. For example, the tensile strengths of all the steels, if stress relieved, are very close to being 500 times the Brinell hardness number when expressed as psi. This is true for both annealed and heat-treated steel. In a following section and in Figure 7-9 (pg. 216), the relationship of the Brinell hardness to the tensile strength is shown to depend upon the strain-hardening or the strain-strengthening exponent. Even though the Brinell hardness test is a technological one, it can be used with considerable success in engineering research on the mechanical properties of materials, and it is a much better test for this purpose than the Rockwell test.

The Brinell hardness number of a given material increases as the applied load is increased, the increase being somewhat proportional to the material's strain-hardening rate. This is due to the fact that the material beneath the indentation is plastically deformed, and the greater the penetration the greater the amount of cold work, with a resulting higher hardness.

Meyer Hardness

Meyer hardness is determined by dividing the load applied to a spherical indenter by the projected area of the indentation. The Meyer hardness test itself is identical to the Brinell and is usually performed on a Brinell hardness machine. The difference between these two hardness scales is simply the area that is divided into the applied load — the projected area being used for the Meyer hardness and the spherical surface area for the Brinell hardness. Both are based on the diameter of the indentation. Meyer hardness is also expressed as kg/mm^2. In this book Meyer hardness is identified by the symbol H_M and is calculated from the equation

$$H_M = \frac{4L}{\pi d^2} \qquad (7\text{-}2)$$

It should be readily apparent that for a given material the Meyer hardness number is greater than the Brinell hardness number.

Because the Meyer hardness is determined from the projected area rather than the contact area, it is a more valid concept of stress and therefore is considered a more basic or scientific hardness scale. Although this is true, the scale has been used very little since first proposed in 1908, and it has been used only in research studies. Its lack of acceptance is probably due to the fact that it does not relate directly to the tensile strength the way that Brinell hardness does.

Strain hardening exponent. Meyer is much better known for the original strain-hardening equation that bears his name than for his hardness scale. The strain-hardening equation for a given diameter of ball is

$$L = Ad^p \qquad (7\text{-}3)$$

where L is the load on a spherical indenter, d is the diameter of the resulting indentation, and p is known as the Meyer strain-hardening exponent. A limited number of values of the strain-hardening exponent for a variety of metals are found in the materials handbooks. They vary from a minimum value of 2.0 for low-work hardening metals to about 2.6 for dead soft brass. The value of p is about 2.25 for both annealed pure aluminum and annealed 1020 steel.

Equation (7.3) plots as a straight line on logarithmic coordinates or when log L is plotted against log d on cartesian coordinates. The exponent p is the slope of the line and A is the value of load corresponding to a diameter of indentation of unity. Figure 7-1 is a plot of experimental load vs. diameter of indentation obtained on an 80%Cu - 20%Zn brass. For this material p is 2.53 and the constant A is 19.3.

Meyer's work also indicated that the strain-hardening exponent p was affected very little by changes in the ball diameter, but that the coefficient A decreased with increasing ball diameter such that the product AD^{p-2} was

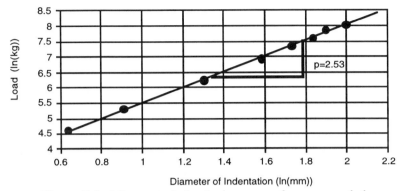

Figure 7-1. Meyer strain-hardening curve for an annealed 80%Cu - 20%Zn brass indented by a 10 mm ball.

approximately equal to a constant B. Thus the strain-hardening equation can be rewritten to include the diameter of the ball as

$$L = \frac{Bd^p}{D^{p-2}} \qquad (7\text{-}4)$$

A very interesting relationship develops when both sides this last equation are divided by d^2, namely

$$\frac{L}{d^2} = B\left(\frac{d}{D}\right)^{p-2} \qquad (7\text{-}5)$$

Equation 7-5 is interesting because it has the same units as the tensile stress-strain relationships. That is, L/d^2 has the unit of stress, kg/mm^2, and d/D has the unit of strain, mm/mm. Experimental data for some metals show that the exponent p–2 in Eq.(7-5), is approximately equal to the strain-strengthening exponent m in the tensile stress-strain equation $\sigma = \sigma_0 \varepsilon^m$ presented later. That is,

$$p\text{--}2 = m \qquad (7\text{-}6)$$

There is no known correlation between the coefficients B and σ_0. In the case of the 80-20 brass that had an experimentally determined value of p of

2.53, a separately run tensile test gave a value of m = 0.53. However, such good agreement does not usually occur, owing partly to the difficulty of accurately measuring the diameter d. Nevertheless, this approximate relationship between the strain-hardening and the strain-strengthening exponents can be very useful in the practical evaluation of a material's mechanical properties. For example, if reasonable care is taken in conducting the tests, both the strain-strengthening exponent m and the actual tensile strength S_u can be determined from a simple hardness test for any metal, not just a steel. An example of this type of calculation is given in the section entitled Tensile Strength and Brinell Hardness Ratio (pg. 214).

Diamond Pyramid or Vickers Hardness

The *diamond pyramid hardness,* or *Vickers hardness* as it is frequently called, is determined by dividing the load applied to a square-based pyramid indenter by the surface area of the indentation. It is similar to the Brinell hardness test except for the indenter used, and is identified in this book as Hp. In other works it is sometimes designated as Hv. The indenter is made of industrial diamond, hence the name, and the two pairs of opposite faces are accurately ground to an included angle of 136°. The loads applied vary from as low as 100 g for microhardness readings to as high as 120 kg for standard hardness readings. The indentation at the surface of the work piece is square shaped. The diamond pyramid hardness number is determined by measuring the length of the two diagonals of the indentation and using the average value in the following equations

$$Hp = \frac{2L\sin(\alpha/2)}{d^2} = \frac{1.8544\,L}{d^2} \qquad (7\text{-}7)$$

where L is the applied load in kilograms, d is the diagonal of the indentation in millimeters, and α is the face angle of the pyramid which is 136°.

A face angle of 136° was selected because it is the angle whose sides are tangent to a circle of diameter D at a chord length d of 0.375 D. This was done to make the pyramid indentation somewhat similar to the Brinell indentation, and it is customary in using the Brinell test to apply loads that give d/D ratios between 0.25 and 0.50, with the average value being 0.375. The pyramid hardness number and the Brinell hardness number for a material are approximately equal when the d/D ratio for the latter is 0.375 or less. But when the d/D ratio exceeds 0.5, the pyramid number is considerably larger than the Brinell.

The main advantage of a cone or pyramid indenter is that it produces indentations that are geometrically similar, regardless of the depth. In order to be geometrically similar the angle subtended by the indentation must be constant, regardless of the depth of the indentation. This is not true of a ball indenter. It is believed that if geometrically similar deformations are produced, then the material being tested is stressed to the same amount, regardless of the depth of the penetration. On this basis it would be expected that a conical or pyramidal indenter would give the same hardness number regardless of the load applied. Experimental data show that the pyramid hardness number is independent of the load if loads greater than 3 kg are applied. However, for loads less than 3 kg, the hardness is affected by the load, depending upon the strain hardening exponent of the material being tested. For example, annealed 17-4 PH stainless steel (p = 2.05) has a hardness of 372Hp with a 3 kg load and only 354Hp with a 100 g load. Thus the pyramid hardness number of a material with a low strain-hardening rate decreases with decreasing loads, whereas the material with a high strain-hardening rate number increases with decreasing load.

This condition makes it possible for one material to be harder than another with one testing load and to be softer at a different testing load. This is true for the two materials discussed here since the hardness values with a 50 g load is 325Hp for the 17-4 PH steel and 360Hp for the cobalt superalloy. The ranking is reversed for 3 kg loads. But for loads greater than 3 kg the pyramid hardness number for a given material is constant.

The Brinell hardness number for all materials increases as the applied load is increased. The amount of increase is very slight for a material having a strain-hardening exponent of 2.00 but is very significant for a material having a p value of 2.6. The cobalt base alloy (p = 2.50) mentioned previously has a hardness of 150 H_B with a 500 kg load and 201 H_B with a 3000 kg load.

Knoop Hardness

The *Knoop hardness* is determined by dividing the applied load to a special rhombic-based pyramid indenter by the projected area of the indentation. It is identified as H_K in this book. The indenter is made of industrial diamond and the four pyramid faces are ground so that one of the angles between the intersections of the four faces is 172.5° and the other angle is 130°. A pyramid of this shape makes an indentation that has the projected shape of a parallelogram having a long diagonal that is 7 times as long as the short diagonal and 30 times as long as the maximum depth of the indentation. It is sometimes referred to as a sensitive pyramidal indenter since its elongated shape results in measurable changes in length of the long diagonal with small loads.

The greatest application of the Knoop hardness is in the microhardness area. As such, the indenter is mounted on an axis parallel to the barrel of a microscope having magnifications of 100× to 1500×. A metallurgically polished, either etched or unetched, flat specimen is used. The location at which the hardness is to be determined is located and repositioned under the hairline of the microscope eyepiece. The specimen is positioned under the indenter and the load is applied for 10 – 20 seconds. The specimen is then located under the microscope again and the length of the long diagonal is measured. The Knoop hardness number is then determined from the following equation (or tables constructed from the equation):

$$H_K = \frac{L}{0.07028\,d^2} \qquad (7\text{-}8)$$

where L is the applied load in kilograms, d is the length of the long diagonal in millimeters, and 0.07028 is the indenter constant for the above-mentioned standard angles.

The Knoop hardness, and also the diamond pyramid hardness, can be used to determine the hardness of individual grains or microconstituents. Loads of 100 g to 3 kg are most common for microhardness studies. Care must be exercised in comparing microhardness values because, as explained in the preceding section, the hardness number for any given material varies with the applied load if loads less than 3 kg are used. This is true for the Knoop as well as the diamond pyramid hardness, although with the Knoop it appears that the hardness of all materials (with low and high strain hardening rates) increases as the load decreases below 3 kg. Thus the paradox of one material being harder than another at a low load but softer with a large load does not occur with the Knoop hardness as it does with the diamond pyramid (Vickers) hardness.

Scleroscope Hardness

The *Scleroscope hardness* is determined by measuring the height to which a special indenter bounces. It is frequently referred to as the shore scleroscope since the only type in use today is the one developed by Shore. The Shore indenter has a rounded end and falls freely a distance of about 10 inches in a glass tube. The rebound height is determined by visually observing the maximum height to which the top of the indenter reaches on a scale attached to the glass tube. The scale is divided into 140 equal divisions and numbered from 0 to 140. The scale was selected so that the rebound height from a fully hardened high carbon steel (65 to 67 R_C or 750 to 800 H_B) gives a maximum reading of 100.

All of the previously described hardness scales are "static" hardnesses since the load is slowly applied and maintained for several seconds. The scleroscope hardness on the other hand is a "dynamic" hardness. As such, it is greatly influenced by the elastic modulus of the material being tested.

The principal advantages of the Shore scleroscope hardness tester are that it is small and easily portable, and leaves a very small indentation. Its main disadvantage is that it is so much influenced by the elastic properties of the material that in some cases softer (as determined by static hardness tests) materials give higher scleroscope readings than harder materials.

THE TENSILE TEST

The tensile test is conducted on a machine that can apply uniaxial tensile or compressive loads to the test specimen; it also has provisions for accurately registering the value of the load and the amount of deformation that occurs to the specimen. The tensile specimen may be a round cylinder or a flat strip and has a reduced cross section, called the gage section, at its mid-length to ensure that the fracture does not occur at the holding grips. The minimum length of the reduced section, the gage length, for a standard specimen is 4 times its diameter. The most commonly used specimen has a 0.505" diameter gage section (0.2 sq. in. cross-sectional area) that is 2 1/4" long to accommodate a 2" long gage section. The overall length of the specimen is 5 1/2" with 1" length of 3/4-10 NC threads on each end. The ASTM specifications list several other standard sizes, including flat specimens.

In addition to the tensile properties of strength, rigidity, and ductility, the tensile test also gives information regarding the stress-strain behavior of the material. It is very important to distinguish between *strength* and *stress* as they relate to material properties and mechanical design. But it is also somewhat awkward to make this distinction since they have the same units and many books use the same symbol (σ) for both.

Strength is a property of a material — it is (1) a measure of the material's ability to withstand stress, or, (2) the material's load-carrying capacity. The numerical value of strength is determined by dividing the appropriate load (yield, maximum, fracture, shear, cyclic, creep, etc.) by the original cross-sectional area of the specimen, and is designated as S. Thus,

$$S = L / A_o \qquad (7\text{-}9)$$

The subscripts y, u, f, and s are appended to S to denote yield, ultimate, fracture, and shear strength, respectively. Although the strength values

188   Chapter 7

obtained from a tensile test have the *units* of stress, psi or the equivalent, they are not really *values* of stress. A part, such as a shaft or a linkage, has the same strength regardless of whether it is lying unassembled and unloaded on a storage shelf or whether it is actually functioning in service.

Stress is a condition of a material due to an applied load. If there are no loads on a part, then there are no stresses in it. (Residual stresses may be considered as being caused by unseen loads.) The numerical value of the stress is determined by dividing the actual load or force on the part by the actual cross-sectional area that is supporting the load. Normal stresses are almost universally designated by the symbol σ and the stresses due to tensile loads are determined from the expression

$$\sigma = L / A_i \tag{7-10}$$

where A_i is the instantaneous cross-sectional area corresponding to that particular load. The units of stress are psi, or the equivalent.

During a tensile test, the stress varies from zero at the very beginning to a maximum value that is equal to the true fracture stress, with an infinite number of stresses in between. However, the tensile test gives only three values of strength: the yield, the ultimate, and the fracture. An appreciation of the differences between strength and stress will be achieved after reading the following sections on the use of tensile test data.

Engineering Stress-Strain (Strength-Nominal Strain)

The tensile test has been traditionally used to determine the so-called engineering stress-strain data that is needed to plot the engineering stress-strain curve for a given material. However, since the engineering stress is not really a stress but is a measure of the material's strength, it is more appropriate to call such data either strength-nominal strain or nominal stress-strain data. Table 7-1 illustrates the data normally collected during a tensile test, and Figure 7-2 shows the condition of a standard tensile specimen at the time the specific data in the table are recorded. The load-vs.-gage length data, usually plotted as an elastic stress-strain curve, are needed to determine Young's modulus of elasticity of the material as well as the proportional limit. These data are also needed to determine the yield strength if the offset method is used. All of the definitions associated with the engineering stress-strain, or more appropriately with the strength-nominal strain properties, are presented in the section entitled "Tensile Properties" (pg. 199) and are discussed in conjunction with the experimental data for commercially pure titanium shown in Table 7-1 and Figure 7-2.

Mechanical Properties

Table 7-1. Tensile Test Data

Material:	A40 Titanium	Condition:	Annealed
Specimen size:	0.505" diameter x 2" gage length		$A_o = 0.200$ in.^{2}
Yield load:	9,040 lbs	Yield Strength:	45.2 ksi
Maximum load:	14,950 lbs	Tensile strength:	74.75 ksi
Fracture load:	11,500 lbs	Fracture strength:	57.5 ksi
Final length:	2.48"	Elongation:	24%
Final diameter:	0.352"	Reduction of Area:	51.15%
Load (lbs)	Gage length (in.)	Load (lbs)	Gage length (in.)
1,000	2.0006	6,000	2.0044
2,000	2.0012	7,000	2.0057
3,000	2.0018	8,000	2.0070
4,000	2.0024	9,000	2.0094
5,000	2.0035	10,000	2.0140

The elastic and elastic-plastic data listed in the table is plotted in Figure 7-3 with an expanded strain axis, which is necessary for the determination of the yield strength. The nominal (approximate) stress, or the strength S, which is calculated from Eq. (7-9) and is plotted as the ordinate, is the load divided by the original area, which, for this standard specimen, is 0.200 sq. in.

The abscissa of the engineering stress-strain plot is the *nominal strain*, which is defined as the unit elongation obtained when the change in length is divided by the original length. Nominal strain is expressed in units of in./in. and in this book is designated as n. Thus for tension

$$n = \frac{\Delta l}{l} = \frac{l_f - l_o}{l_o} \qquad (7\text{-}11)$$

or

$$n = \frac{\Delta A}{A} = \frac{A_o - A_f}{A_f} \qquad (7\text{-}12)$$

where l is the gage length, A is the cross-sectional area, and the subscripts o and f designate the original and final state, respectively. Both equations give the same numerical value of nominal strain if the deformation is uniform along the entire gage length; that is, if no "neck" is present. This is true for all materials

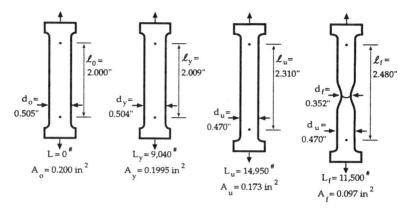

Figure 7-2. A standard tensile specimen of A40 titanium at various stages of loading: (a) unloaded, (b) at the yield load, (c) at the maximum load, (d) at the fracture load.

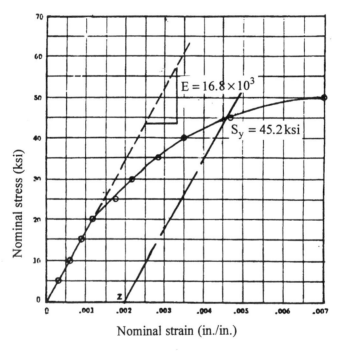

Figure 7-3. The elastic-plastic portion of the engineering stress-strain curve for annealed A40 titanium.

with deformation strains that do not exceed the strain at the maximum load of a tensile specimen.

The data for annealed titanium in Figure 7-2 show a nominal strain at the time that the maximum load is on the specimen:

from Eq. (7-11) $n_u = (2.310 - 2.000)/2.000 = 0.310/2.000 = 0.155$ in./in.
from Eq. (7-12) $n_u = (0.200 - 0.173)/0.173 = 0.027/0.173 = 0.155$ in./in.

However, after nonuniform deformation (necking) occurs, the nominal strain on the basis of length (Eq. (7-11)) has no real meaning, but the nominal strain on the basis of the minimum area (Eq. (7-12)) does have real meaning. For the titanium data in Figure 7-2, the nominal strain at fracture is:

from Eq. (7-11) $n_f = (2.480 - 2.000)/2.000 = 0.480/2.000 = 0.24$ in./in.
from Eq. (7-12) $n_f = (0.200 - 0.097)/0.097 = 0.103/0.097 = 1.06$ in./in.

In this case the value given by Eq. (7-11) is an average value over the entire gage section whereas the value from Eq. (7-12) is the actual strain that occurs at the root of the neck of the specimen and is a valid strain value for engineering calculations.

It is customary to plot the data obtained from a tensile test as a stress-strain curve such as that illustrated in Figure 7-4 but without including the word "nominal." The reader may be tempted to consider such a curve an actual stress-strain curve, which it obviously is not. The curve plotted in Figure 7-4 is in reality a load-deformation curve. If it were so identified in the introductory texts on material properties and if the ordinate axis were labeled load (lbs) rather than stress (psi), the distinction between strength and stress would be more apparent. Although the fracture load is less than the ultimate load, the stress in the material just prior to fracture is much greater than the stress at the time the ultimate load is applied to the specimen.

True Stress-Strain

The tensile test is also used to obtain true stress-strain, or true stress-natural strain data to define a material's plastic stress-strain characteristics. It is ironic that "true" is employed only to distinguish a "true" stress from an "engineering" stress. This distinction would not be necessary if curves such as that shown in Figure 7-4 were identified as load-deformation curves and if the values S_u and

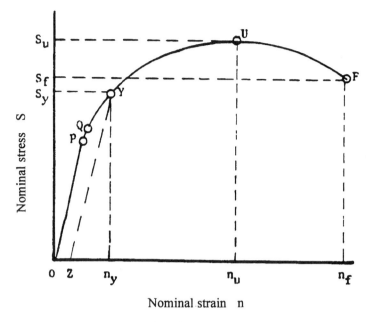

Figure 7-4. The engineering stress-strain curve. P is the proportional limit, Q is the elastic limit, Y is the yield load, U is the ultimate (maximum) load, F is the fracture l load.

S_f referred to strength rather than engineering stress. In this situation there would then be only one type of stress, which in reality is the case.

To obtain stress-strain data in the plastic deformation region, it is necessary to measure and record simultaneously the cross-sectional area of the specimen and the load on it. For round sections, it is sufficient to measure the diameter, or if the circular section becomes elliptical, to measure the minimum and maximum diameters and record the average value. The load-deformation data in the plastic region of the tensile test of an annealed titanium are listed in Table 7-2. These data are from a continuation of the tensile test yielding the elastic data shown is given in Table 7-1. During the elastic deformation portion of the test, the gage length is measured with a sensitive extensometer to determine the strain. During the plastic deformation portion of the test, the extensometer is removed and the minimum diameter of the specimen is measured with a micrometer, or a similar instrument, to determine the strain. This method avoids the problem encountered in measuring the "average" linear strain on specimens that include a necked region. The load-diameter data in

Table 7-2 are recorded during the test and the remainder of the table is completed afterwards. The values of stress are calculated by means of Eq. (7-10), $\sigma = L / A_i$. The strain in this case is the natural strain or logarithmic strain, which is the sum of all the infinitesimal nominal strains or unit elongations. That is

$$\varepsilon = \frac{\Delta l_1}{l_o} + \frac{\Delta l_2}{l_o + \Delta l_1} + \frac{\Delta l_3}{l_o + \Delta l_1 + \Delta l_2} + \ldots$$

and in integral form

$$\varepsilon = \int_{l_o}^{l_f} \frac{dl}{l}$$

or

$$\varepsilon = \ln \frac{l_f}{l_o} \tag{7-13}$$

Compressive strains are considered by definition to be negative so, for compressive deformation, the strain is calculated from

$$\varepsilon = \ln \frac{l_o}{l_f} \tag{7-14}$$

The negative sign can be disregarded if the strain is defined as

$$\varepsilon = \ln \frac{l_l}{l_s} \tag{7-15}$$

where the subscripts l and s refer to the larger and smaller lengths, respectively.

The volume of material remains constant during plastic deformation. That is,

$$V_o = V_f \quad \text{or} \quad A_o l_o = A_f l_f$$

Thus, for tensile deformation, Eq. (7-13) can be expressed as

$$\varepsilon = \ln \frac{A_o}{A_f} \tag{7-16}$$

Chapter 7

Table 7-2. Load Deformation Data From Tensile Test
(see Table 7-1)

Load (lbs)	Diam. (in.)	Area (sq. in.)	Area ratio	Stress (psi)	Strain (in./in.)
12,000	0.501	0.197	1.015	60.9	0.0149
14,000	0.493	0.191	1.048	73.5	0.0473
14,500	0.486	0.186	1.075	78.0	0.0724
14,950	0.470	0.173	1.155	87.5	0.144
14,500	0.442	0.153	1.308	94.8	0.268
14,000	0.425	0.142	1.410	99.4	0.344
11,500	0.352	0.097	2.06	119.0	0.720

For compressive deformation, the strain is determined from the cross-sectional areas by the expression

$$\varepsilon = -\ln\frac{A_f}{A_o} = \ln\frac{A_o}{A_f} \qquad (7\text{-}17)$$

In the general case where the negative sign can be ignored

$$\varepsilon = \ln\frac{A_l}{A_s} \qquad (7\text{-}18)$$

where the subscripts l and s refer to the large area and the small area. respectively.

Quite frequently in calculating the strength or the ductility of a cold-worked material it is necessary to determine the value of strain ε that is equivalent to the amount of the cold work. The amount of cold work is defined as the percent reduction of a material's cross-sectional area (or simply the percent reduction of area) by a plastic deformation process. It is designated by the symbol W and is determined from the expression

$$W = \frac{A_l - A_s}{A_l} \times 100 \qquad (7\text{-}19)$$

where the subscripts l and s refer to the large and the small areas, respectively. Eq. (7-19) can be expressed as a ratio of A_l / A_s as follows

$$\frac{A_1}{A_s} = \frac{100}{100-W} \tag{7-20}$$

If this ratio is substituted into Eq. (7-18), the appropriate relationship between strain and cold work results:

$$\varepsilon_w = \ln\left(\frac{100}{100-W}\right) \tag{7-21}$$

The subscript w identifies this strain as that associated with the percent cold work W.

The stress-strain data of Table 7-2 are plotted in Figure 7-5 on cartesian coordinates. The most significant difference between the shape of this stress-strain curve and that of the load-deformation curve of Figure 7-4 is that the stress continues to rise until fracture occurs and does not reach a maximum value as the load does. As can be seen in Table 7-2 and Figure 7-5, the stress at the time of the maximum load is 86 ksi and it increases to 119 ksi at the instant that fracture occurs. Although a smooth curve can be drawn through the experimental data, it is not a straight line and consequently many experimental points are necessary to accurately determine the shape and position of the curve. The stress-strain curve in Figure 7-5 becomes a straight line when plotted on logarithmic coordinates, as is true for the plotting of stress-strain curves for most engineering metals in the range of strains between about 2% and the maximum load strain.

The stress-strain data obtained from the tensile test of the annealed A40 titanium listed in Tables 7-1 and 7-2 are plotted on logarithmic coordinates in Figure 7-6. The elastic portion of the stress-strain curve is also a straight line on logarithmic coordinates, as it is on cartesian coordinates. This is true for all materials, not just the titanium alloy plotted here. When plotted on cartesian coordinates, the slope of the elastic modulus is different for all materials. However, when plotted on logarithmic coordinates, the slope of the elastic modulus is one (unity) for all materials — it is only the height, or position, of the line that is different for all materials. In other words, the elastic moduli for all materials are parallel lines making an angle of 45° with the ordinate axis. The modulus line for a given material goes through a point that corresponds to the intersection of a line drawn vertically from a strain ε of 1 and the line drawn horizontally from a stress σ that is numerically equal to E, Young's modulus. That is, E is the value of true stress on the extrapolated elastic stress-strain curve corresponding to a strain of 1. This can be seen from the equation for the elastic modulus of Hooke's law, which is

196   Chapter 7

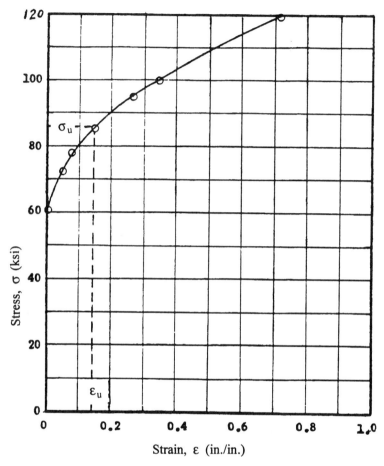

Figure 7-5. Stress-strain curve for annealed A40 titanium. The strain is the natural, or logarithmic, strain and the data of Tables 7-1 and 7-2 are plotted on cartesian coordinates.

$$\sigma = E\varepsilon \qquad (7\text{-}22)$$

where the exponent for ε is one and therefore is not written. In logarithmic form, the equation is

$$\log \sigma = \log E + 1 \log \varepsilon$$

Mechanical Properties 197

which is of the form

$$y = b + mx$$

where m (or 1) is the slope of the straight line. The modulus of elasticity E for this titanium alloy is equal to 16.8×10^6 psi.

The experimental points in Figure 7-6 for strains greater than 0.01 (1% plastic deformation) also fall on a straight line having a slope of 0.14. The slope of the stress-strain curve on logarithmic coordinates is called the *strain-strengthening exponent* because it indicates the increase in strength that results from plastic strain. It is also referred to as the strain-hardening exponent, which is somewhat misleading since the real strain-hardening exponent is the Meyer exponent, discussed previously. In this book the strain-strengthening exponent is represented by the symbol m and the strain-hardening exponent by the symbol p. To a first approximation, the two are related by the expression $p = m + 2$ (Eq. (7-6)).

The equation for the plastic stress-strain line is

$$\sigma = \sigma_o \varepsilon^m \qquad (7\text{-}23)$$

and is known as the strain-strengthening equation since σ is directly related to the yield strength. As will be shown later, in the development of Eq. (7-41), Eq. (7-23) is the locus of yield strengths for cold-worked metals where ε defines

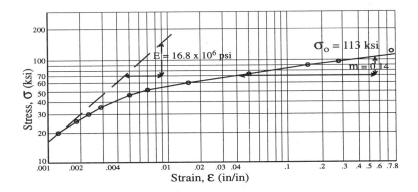

Figure 7-6. Stress-strain curve for annealed A40 titanium plotted on logarithmic coordinates. The data are the same as in Figure 7-5.

the amount of the prior cold work that was given to the material. However, this is true only for yield strengths having the same sense and direction as the cold work strain.

The proportionality constant σ_0 is called the strength coefficient in this book. In other books it is simply called a constant and is identified by the symbol k. A material property that is similar to the elastic constant E is too important a constant, in the author's belief, to be designated by the general symbol k. In fact, σ_0 is to a material's plastic behavior what E is to its elastic behavior. E is the value of stress associated with an elastic strain of one; σ_0 is the value of stress associated with a plastic strain of one. This follows from the fact that one raised to any power is still one, so then $\sigma = \sigma_0$ when $\varepsilon = 1$.

The amount of cold work necessary to give a strain of unity is determined from Eq. (7-21) to be 63.2%. Since the stress-strain curve is the locus of yield strengths for the corresponding amount of prior strain, σ_0 can be defined as the yield strength of the material after it as been given 63.2% cold work. The value of σ_0 for the titanium data plotted in Figure 7-6 is 113 ksi. Thus if this annealed titanium alloy were given 63.2% cold work by rolling, for example, the longitudinal tensile yield strength of the cold-rolled metal would be 113 ksi whereas the annealed yield strength is only 45.2 ksi and the tensile strength of the annealed material is only 75.75 ksi. The values for the strength coefficient σ_0 and strain-strengthening exponent m for a large variety of materials and several different heat treatments are listed in Table B-2 of Appendix B (p. 341).

For most materials there is an elastic-plastic region between the two straight lines of the fully elastic and the fully plastic portions of the stress-strain curve. A material that has no elastic-plastic region may be considered an "ideal" material because the study and analysis of its tensile properties are simpler. Such a material has a complete stress-strain relationship that can be characterized by two intersecting straight lines, one for the elastic region and one for the plastic region. Such a material would have a stress-strain curve similar to the one labeled I in Figure 7-7. A few real materials have a stress-strain curve that approximates the "ideal" curve. However, most engineering materials have a stress-strain curve that resembles curve O in Figure 7-7. These materials appear to "over-yield." That is, they have a higher yield strength than the "ideal" value, followed by a region of low or no strain strengthening before the fully plastic region begins. Among the materials that have this type of curve are steel, stainless steel, copper, brass alloys, nickel alloys, and cobalt alloys.

Only a few materials have a stress-strain curve similar to that labeled U in Figure 7-7. The characteristic feature of this type of material is that it appears to "under-yield" — that is, it has a yield strength that is lower than the "ideal" value. Some fully annealed aluminum alloys have this type of curve.

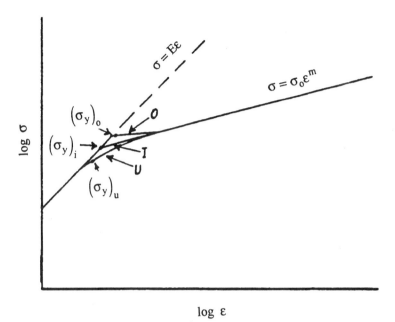

Figure 7-7. Schematic representation of three types of stress-strain curves. I is an "ideal" curve, and U and O are the two types of real curves.

TENSILE PROPERTIES

Tensile properties, those mechanical properties determined by a tension test, are used more frequently than any other of the mechanical properties as the basis of mechanical design of structural components. Tensile data are available for materials more than any other type of material property data. Frequently the design engineer has to base mechanical property calculations on the tensile properties — even under cyclic, shear, or impact loading — simply because more appropriate mechanical property data are not available for the material being considered for a specific part. As Professor Joseph Shigley states in his book *Mechanical Engineering Design*," the fundamental problem of the designer is to use the simple tension-test data and relate them to the strength of the part, regardless of the stress state or the loading situation." All of the tensile properties are defined in this section; each is briefly discussed on the basis of the tensile test described in the preceding section.

Modulus of Elasticity

The *modulus of elasticity* is the ratio of stress to the corresponding strain during elastic deformation. It is the slope of the straight line (elastic) portion of the stress-strain curve when drawn on cartesian coordinates. It is also known as Young's modulus, or the proportionality constant in Hooke's law, and is commonly designated as E with the unit of psi, ksi or MPa. The modulus of elasticity of the titanium alloy whose tensile test data is reported in Table 7-1 is shown in Figure 7-3, where the first four experimental data points fall on a straight line having a slope of 16.8×10^6 psi.

Proportional Limit

The *proportional limit* is the greatest stress which a material is capable of withstanding without any deviation from a linear proportionality of stress to strain. It is the point where a straight line drawn through the experimental data points in the elastic region first departs from the actual stress-strain curve. Point P in Figure 7-4 is the proportional limit (20 ksi) for this titanium alloy. The proportional limit is very seldom used in engineering specifications because its determination depends so much upon the sensitivity and accuracy of the testing equipment and the subjective judgment of the person evaluating the data.

Elastic Limit

The *elastic limit* is the greatest stress which a material is capable of withstanding without any permanent deformation after removal of the load. It is designated as point Q in Figure 7-4. The elastic limit is also very seldom used in engineering specifications because of the complex testing procedure of many successive loadings and unloadings that is necessary for its determination.

Yield Strength

The *yield strength* is the nominal stress at which a material undergoes a specified permanent deformation. There are several methods to determine the yield strength. For materials such as low-carbon steel that have a yield point phenomenon as described below, the "drop of the beam" method is simple and reliable. In this method a moderate rate of loading is selected and not changed during the early part of the test. The load-indicating needle then rotates at a fixed speed until it reaches the upper yield point, at which time the load actually decreases or drops and the indicating needle reverses direction for a short time. The point at which the load drops off is reported as the yield load and is used to calculate the yield strength. A variation of this method is used for materials that do not have a yield point phenomenon where the load actually drops off. In this case the operator adjusts the loading rate so that the load-indicating needle rotates at the same speed that a pacing disc mounted under the needle is rotating. The operator then records the yield load as the load at which the indicating needle fails to keep up with the pacing disc.

The divider method is very simple and gives results which are satisfactory for most engineering work except for research on mechanical properties. In this method, the operator adjusts the legs of a pair of dividers so that their ends fall into the punch marks that define the gage length of the specimen. The divider ends are periodically placed or "felt" into the punched holes while the load is being applied to the specimen. The load that is on the specimen at the first time when the divider ends do not fit the punch mark is recorded as the yield load.

The most reliable and consistent method of determining the yield strength is the offset method. This method requires that the nominal stress-strain diagram be first drawn on cartesian coordinates. A point Z is placed along the strain axis at a specified distance from the origin as shown in Figures 7-3 and 7-4. A line parallel to the elastic modulus is drawn from Z until it intersects the nominal stress-strain curve. The value of stress corresponding to this intersection is called the yield strength by the offset method. The distance OZ is called the offset and is expressed as a percentage. The most common offset is 0.2% which corresponds to a nominal strain of 0.002 in./in. This is the value of offset used in Figure 7-3 to determine the yield strength of the A40 titanium. An offset of 0.01% is sometimes used, and the corresponding nominal stress is called the proof strength, which is a value very close to the proportional limit.

For some nonferrous materials, an offset of 0.5% is used to determine the yield strength.

Inasmuch as all of the above methods of determining the yield strength give somewhat different values for the same material it is important to specify what method, or what offset, was used in conducting the test.

Yield Point

The *yield point* is the first stress below the maximum load at which an increase in strain occurs without an increase in stress. This means there is a horizontal portion (or a portion with a negative slope) of the nominal stress-strain curve at the yield strength. The highest stress reached before the load drops off is called the *upper yield point*. The lowest value that it drops to before rising again is called the *lower yield point*. Soft low-carbon steels are the only common structural materials that exhibit a pronounced yield point.

Tensile Strength

The *tensile strength* is the value of nominal stress obtained when the maximum (or ultimate) load that the tensile specimen supports is divided by the original cross-sectional area of the specimen. It is shown as S_u in Figure 7-4 and is sometimes called the ultimate strength. The tensile strength is a commonly used mechanical property in engineering calculations even though it is the yield strength that is used to measure when plastic deformation begins for a given material. The real significance of the tensile strength, as a material property, is that it indicates what maximum load a given part can carry in uniaxial tension without breaking. It determines the absolute maximum limit of load that a part can support.

Fracture Strength

The *fracture strength* or *breaking strength* is the value of nominal stress obtained when the load carried by a tensile specimen at the time of fracture is divided by its original cross-sectional area. The breaking strength is not used as a material property in mechanical design.

Only the three strengths described above (yield strength, tensile strength, and fracture strength) are determined by a tensile test. Although they are usually represented on an engineering stress-strain curve, as shown in Figure 7-4, it must be kept in mind that such a curve is only schematic, because the strains at the ultimate and fracture points are not known.

Reduction of Area

The *reduction of area* is the maximum change in area of a tensile specimen divided by the original area and is usually expressed as a percentage. In this book it is designated as A_r and is calculated as follows:

$$A_r = \frac{A_o - A_f}{A_o} \times 100 \qquad (7\text{-}24)$$

where the subscripts o and f refer to the original and final areas, respectively. The reduction of area, along with the strain at maximum load n_u, is the best measure of a material's ductility. An equivalent, but simpler to use, measure of ductility is the area ratio R, which is the ratio of the original cross-sectional area of the tensile specimen to the final minimum area of the fractured specimen. That is: $R = A_o / A_f$.

Fracture Strain

The *fracture strain* is the true strain at fracture of the tensile specimen. It is represented by the symbol ε_f and is calculated from the definition of strain given in Eq. (7-16), which is $\varepsilon_f = \ln(A_o / A_f)$. This area strain must be used rather than the lineal strain given in Eq. (7-16), which is $\varepsilon = \ln(l_f / l_o)$, whenever the tensile specimen exhibits a neck, because the latter expression gives an average strain rather than the fracture strain. If the percent reduction of area A_r is known for a material, the fracture strain can be calculated from the expression

$$\varepsilon_f = \ln(\frac{100}{100 - A_r}) \tag{7-25}$$

This expression is arrived at by rewriting Eq. (7-24) as

$$\frac{A_o}{A_f} = \frac{100}{100 - A_r} \tag{7-26}$$

and substituting it into Eq. (7-16).

Percent Elongation

The *percent elongation* is a crude measure of a material's ductility obtained when the change in gage length of a fractured tensile specimen is divided by the original gage length and expressed as a percent. It is customarily designated as El. Thus

$$El = \frac{l_f - l_o}{l_o} \times 100 \tag{7-27}$$

Since most materials exhibit non-uniform deformation before fracture occurs during a tensile test, the percent elongation is only an average value and as such cannot be used in meaningful engineering calculation.

The percent elongation is not really a material property, but rather a combination of a material property and a test condition. A true material property is not significantly affected by size of the specimen. Thus a 1/4" diameter and a 1/2" diameter tensile specimen of the same material give the same values for yield strength, tensile strength, reduction of area or fracture strain, modulus of elasticity, strain-strengthening exponent and strength coefficient. But a 1" gage length specimen and a 2" gage specimen of the same material do not give the same percent elongation. In fact, for specimens of the same diameter, the percent elongation for a 1" gage length specimen may be 100% greater than that for a 2" gage length specimen.

Strain-Strengthening Exponent

The *strain-strengthening exponent* is the tensile property that indicates the rate at which the yield strength of a material increases as a function of induced ulastic (cold work) deformation (refer to the section "True Stress-Strain," p. 191). It is a basic material property.

Stress Coefficient

The *stress coefficient* is the plastic modulus of a material. It is the proportionality constant in the exponential form of the plastic stress-strain relationship $\sigma = \sigma_o \varepsilon^m$ of a material (refer to the section "True Stress-Strain," p. 191).

STRENGTH-STRESS-STRAIN RELATIONSHIPS

The following relationships among strength, stress, and strain help one to specify the best material for a structural part and to recognize potential fabrication problems by revealing what can be expected of a material during a manufacturing process. Such an understanding also enables an engineer to more readily determine the mechanical properties of a fabricated part on the basis of the material's original properties and the mechanisms involved with the particular process which is used

Nominal Strain and Area

The relationships between nominal strain and area strain are valid *only when the deformation in the entire gage length is uniform* — when no necking occurs. It is beneficial to define a new tensile property at this time. The *uniform elongation*, Ul, is the equivalent lineal strain associated with the percent reduction of area of a tensile specimen. It is the elongation that would occur in a tensile specimen if the entire gage length reduced uniformly to the minimum diameter at the neck. The uniform elongation is equal to the percent elongation when the deformation is uniform.

Two symbols are used to designate the amount of cold work in terms of the area reduction during plastic deformation. W is the amount of cold work given a material or a part during plastic deformation and is expressed as the percent reduction of cross-sectional area that results from the deformation. Thus

$$W = \frac{A_o - A_f}{A_o} \times 100 \tag{7-28}$$

where the subscript f refers to the final area after deformation. The symbol w is used to denote the fraction cold work, which is the percent cold work divided by 100, thus $w = W/100$.

The amount of cold work W should not be confused with the percent reduction of area A_r even though the Eqs. (7-24) and (7-28) look exactly alike. The symbol A_r designates a tensile property of a material and it refers only to the maximum area reduction of a tensile specimen after fracture. The difference between Eqs. (7-24) and (7-28) is that the subscript f refers to the *fracture* area in the former and the *final* area in the latter equation.

The relationships between the nominal strain and percent cold work can be derived from the definition of nominal strain, Eq. (7-11):

$$n = \frac{l_f - l_o}{l_o}$$

which can be rewritten as

$$l_f / l_o = n + 1$$

The volume of metal remains constant during plastic deformation; therefore,

$$A_o l_o = A_f l_f \quad \text{or} \quad l_f / l_o = A_o / A_f = R$$

From the above expressions we get

$$n = R - 1 \tag{7-29}$$

The definition of cold work given in Eq. (7-28) can be rewritten as

$$\frac{A_o}{A_f} = \frac{100}{100 - W}$$

which is also equal to R. Substituting this last equation into Eq. (7-29) gives

$$n = \frac{100}{100 - W} - 1 \tag{7-30}$$

It must be remembered that the values of n calculated from Eq. (7-30) are valid only if the deformation is uniform over the entire length of the part.

Table C.1 in Appendix C is a table of conversions from one type of deformation specification to another. Thus a 50% reduction of area (W) is equivalent to a uniform elongation of 100%. In some commercial forming operations, such as drawing or extruding, ductile materials are given cold reductions in excess of 90%. In these cases the uniform elongations are greater than 1,000%.

Natural Strain and Nominal Strain

The relationship between these two strains is derived from their definitions. The expression for the natural strain is $\varepsilon = \ln(l_f / l_o)$ The expression for the nominal strain (Eq. (7-11)) can be rewritten as $l_f / l_o = n + 1$. When the latter is substituted into the former, the relationship between the two strains is found to be

$$\varepsilon = \ln(n + 1) \tag{7-31}$$

This equation can also be written as

$$n = e^\varepsilon - 1 \tag{7-32}$$

For small values of strain (10^{-2} in./in. or less) the natural strain equals the nominal strain. This comes from the fact that for small n's ln (n + 1) = n, so $\varepsilon = n$. For example, when $n = 0.002$ then $\varepsilon = \ln 1.002 = 0.001998$. But when $n = 0.200$, $\varepsilon = \ln 1.200 = 0.182$.

True Stress and Nominal Stress

The definition of true stress is $\sigma = L/A_f$. From the constancy of volume: $A_f = A_o(l_o/l_f)$ so that

$$\sigma = \frac{L}{A_o} \cdot \frac{l_f}{l_o}$$

or

$$\sigma = S(n+1) \quad \text{or} \quad \sigma = Se^{\varepsilon} \tag{7-33}$$

Thus at the yield load

$$\sigma_y = S_y(n_y + 1) = S_y(e)^{\varepsilon_y}$$

And at the maximum load

$$\sigma_u = S_u(n_u + 1) = S_u(e)^{\varepsilon_u}$$

Strain-Strengthening Exponent and Maximum Load Strain

One of the most useful of the strength-stress-strain relationships is the one between the strain-strengthening exponent and the strain at maximum load. It is also the simplest since the two are numerically equal, that is $m = \varepsilon_u$. This relationship was first derived by Nadai in 1935 on the basis of the load-deformation curve shown in Figure 7-8. The load at any point along this curve is equal to the product of the true stress on the specimen and the corresponding area. Thus

$$L = \sigma A$$

Also

$$\sigma = \sigma_o \varepsilon^m$$

so therefore

$$L = (\sigma_o \varepsilon^m) A$$

208   Chapter 7

Since $\quad \varepsilon = \ln(A_o / A) \quad$ or $\quad A = A_o / e^{\varepsilon}$

the load-strain relationship can be written as

$$L = \sigma_o A_o \varepsilon^m e^{-\varepsilon}$$

The load-deformation curve shown in Figure 7-8 has a maximum, or zero slope, point on it. Differentiating the last equation and equating the result to zero defines the strain associated with the maximum load. This is done as follows:

$$\frac{dL}{d\varepsilon} = -\sigma_o A_o (\varepsilon^m)(e^{-\varepsilon}) + \sigma_o A_o (e^{-\varepsilon}) m(\varepsilon^{m-1}) = 0$$

Dividing by $\sigma_o A_o e^{-\varepsilon}$ reduces the equation to $(\varepsilon_u)^m = m(\varepsilon_u)^{m-1}$, since the strain at maximum load is designated as ε_u. Dividing by $(\varepsilon_u)^m$ gives

$$\varepsilon_u = m \qquad (7\text{-}34)$$

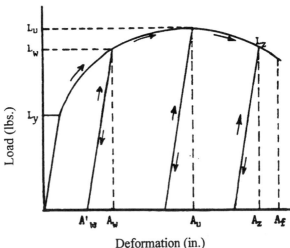

Figure 7-8. A typical load-deformation curve showing unloading and reloading cycles.

Tensile Strength, Stress Coefficient, and Strain-Strengthening Exponent

The following relationship between S_u, σ_o, and m was developed by the author and Professor William Mitchell in the early 1950's. It is a very useful relationship that should be used when reducing any tensile test data to ensure that the calculated values are correct. The author has found many cases of published experimental tensile data that had serious errors, up to 100%, in the reported values because the researchers had not applied this important relationship. The derivation of it is performed below.

By solving Eq. (7-23), $\sigma = \sigma_o \varepsilon^m$, and Eq. (7-33), $\sigma = Se^\varepsilon$, for the stress and strain at the ultimate load, we get

$$\sigma_u = \sigma_o \varepsilon_u^m \quad \text{and}$$
$$\sigma_u = S_u e^{\varepsilon_u} \quad \text{or} \quad \sigma_u = S_u e^m$$

By combining the two new equations we get

$$S_u = \sigma_o (m/e)^m \tag{7-35}$$

Yield Strength and Percent Cold Work

The stress-strain characteristics of a material obtained from a tensile test are shown in Figure 7-6. In the region of plastic deformation the relationship between stress and strain for most materials can be approximated by the equation $\sigma = \sigma_o \varepsilon^m$. When a load applied to a tensile specimen causes a given amount of cold work W (which is a plastic strain of ε_w), the stress on the specimen at that time is σ_w and is defined as

$$\sigma_w = \sigma_o (\varepsilon_w)^m \tag{7-36}$$

Of course σ_w is also equal to the applied load, L_w, divided by the actual cross-sectional area of the specimen, A_w. That is,

$$\sigma_w = L_w / A_w \tag{7-37}$$

210   Chapter 7

If the above tensile specimen were immediately unloaded after reaching L_w, the cross sectional area would increase to A'_w from A_w due to the elastic recovery, or springback, that occurs when the load is removed. This elastic recovery is insignificant for engineering calculations in regard to the strength or stresses on the part. For example, the longitudinal elastic strain associated with yielding stresses (σ_y / E) is of the order of 0.001 to 0.002 in./in. But, since Poisson's ratio is about 0.3, the transverse strain is only 0.0003 to 0.0006 in./in. Thus the unloaded diameter is 1.0005 times the loaded diameter using a strain of 0.0005 in./in.) or the unloaded area $A'_w = (1.0005)^2 A_w = 1.001 A_w$. The unloaded area is 0.1% larger than the loaded area, which is a negligible variation.

If the tensile specimen that has been stretched to a cross sectional area of A'_w is now reloaded, it will deform elastically until the load L_w is again reached. As the load is increased above L_w, the specimen will again deform plastically. This unloading-reloading cycle is shown graphically in Figure 7-8. The yield load for this previously cold-worked specimen is L_w, and the cross-sectional area of the specimen before the reloading is A'_w. Therefore, the yield strength of the previously cold-worked (stretched) specimen is

$$\left(S_y\right)_w = L_w / A'_w \tag{7-38}$$

But since $A'_w = A_w$, then

$$\left(S_y\right)_w = L_w / A_w \tag{7-39}$$

By comparing Eq. (7-39) and (7-37) it is apparent that

$$\left(S_y\right)_w \approx \sigma_w \tag{7-40}$$

And by substituting this last relationship into Eq. (7-36) we get

$$\left(S_y\right)_w \approx \sigma_o\left(\varepsilon_w\right)^m \tag{7-41}$$

Thus it is apparent that *the ulastic (cold-worked) portion of the $\sigma - \varepsilon$ curve is also the locus of yield strengths for a material as a function of the*

Mechanical Properties   211

amount of cold work. This is true for tensile yield strength after tensile deformation, and compressive yield strength after compressive deformation.

Tensile Strength and Percent Cold Work

It is believed by most materials and mechanical design engineers that the relationships between the tensile strength of a cold-worked material and the amount of cold-work given it are the experimentally determined tables and graphs that are provided by material manufacturers, and that the results are different for each family of materials. However, on the basis of the concepts of the tensile test presented earlier in this chapter, the author and Mitchell have derived two relationships between tensile strength and percent cold work that are valid when the prior cold work is tensile. These relationships are derived on the basis of load-deformation characteristics of a material, as represented in Figure 7-8.

Consider the tensile tests of two standard 0.505" diameter specimens of annealed AISI 1020 steel. The first specimen is given the traditional tensile test, and the following results as portrayed in Figure 7-8 are obtained: yield load = 7,000 lbs; ultimate load = 12,000 lbs; fracture load = 8,000 lbs. The second specimen is then given an interrupted tensile test. That is, it is first loaded to L_w which can be any load between the yield and the ultimate load. For this discussion L_w is 9,840 lbs. The diameter of the specimen is 0.492" and the area is 0.190 sq. in. while the load is still applied. The load is rapidly released after reaching 9,840 lbs and the specimen "springs back" elastically. The longitudinal elastic strain is determined from Hooke's law as the stress divided by the elastic modulus, or $51,700 \div 30 \times 10^6$ which is 0.00172 in./in. The diameter increases by Poisson's elastic ratio, which is about 0.3. Therefore the increase of the diameter is 0.492 × 0.3 × 0.00172, or 0.00025." And the true final unloaded diameter is 0.49225." Since the limit of accuracy in reading a standard micrometer is about 0.0002," the unloaded diameter would probably be recorded as simply 0.492." But even if the true diameter is recorded, its corresponding area is 0.1902 sq. in. which is only 0.1% larger than the as-loaded area of 0.190 sq. in. Thus, it can be said that the load area A_w and the unloaded area A'_w are equal.

Now consider the results that are obtained when a tensile test is conducted on this unloaded specimen. Its original diameter is now recorded as 0.492" and its area, or A_w, is 0.190 sq. in. In this case $A_w = 0.95 A_o$, or, in other words, the first loading was such that the specimen was given 5% cold work (5% reduction in cross-sectional area). Upon reloading, this specimen now deforms elastically until reaching the previous high load L_w (9,840 lbs) whereupon it again begins to deform plastically, as illustrated in Figure 7-8. The yield strength of this previously deformed specimen, designated as $(S_y)_w$ is 9,840 lbs ÷ 0.190 sq. in. or 51,700 psi. The load-deformation curve for this prestrained specimen after yielding is identical to the load-deformation curve beyond L_w of the first specimen that was given the uninterrupted tensile test. The maximum load that this prestrained specimen achieves is also L_u, or 12,000 lbs. Therefore the tensile strength of this specimen is determined as follows:

$$S_u = L_u / A'_w = 12,000 / 0.190 = 63,200 \text{ psi}$$

On the basis of this model of the load-deformation characteristics of solid materials, the following general relationship between the tensile strength of a cold-worked material, designated as $(S_u)_w$, and the per cent cold work W can be derived. For any specimen that is given a tensile deformation such that A_w is equal to or less than A_u, by definition

$$(S_u)_w = L_u / A'_w$$

And also by definition

$$L_u = (S_u)_o \times A_o$$

where $(S_u)_o$ is the tensile strength of the original non-cold-worked specimen and A_o is its original area.

The per cent work associated with the deformation of the specimen from A_o to A'_w is

$$W = \frac{A_o - A'_w}{A_o} \times 100 \quad \text{or} \quad w = \frac{A_o - A'_w}{A_o}$$

where w is equal to the percent cold work divided by 100. Thus

$$A'_w = A_o(1-w)$$

By substitution into the first equation

$$(S_u)_w = \frac{(S_u)_o \times A_o}{A_o(1-w)} = \frac{(S_u)_o}{1-w} \quad \text{or} \quad = (S_u)_o e^\varepsilon \quad (7\text{-}42)$$

Thus *the tensile strength of a material that is prestrained in tension to a strain less than its ultimate load strain is equal to its original tensile strength divided by one minus the fraction cold work.* This relationship is valid for any deformations less than the deformation associated with the ultimate load. That is, for $A_w \le A_u$ or $\varepsilon_w \le \varepsilon_u$.

Another relationship can be derived for the tensile strength of a material that has been previously cold-worked in tension by an amount greater than the deformation associated with the ultimate load. This analysis is again made on the basis of Figure 7-8. Consider another standard tensile specimen of 1020 steel that is loaded beyond L_u (12,000 lbs) to some load L_z, say 10,000 lbs. If dead weights were placed on the end of the specimen it would break catastrophically when the 12,000 pound load is applied. But if the load is applied by means of a mechanical screw or a hydraulic pump, then the load will drop off slowly as the specimen is stretched. For this particular example the load is instantly removed when it drops to L_z or 10,000 lbs. The unloaded specimen is not broken, although it may have a "necked" region and it has a minimum cross-sectional area A'_z of 0.101 sq. in. and a diameter of 0.358 in. Now when this same specimen is again loaded in tension it deforms elastically until the load reaches L_z (10,000 lbs) and then it deforms plastically. But L_z is also the maximum value of load that this specimen reaches on reloading. It never again supports a load of L_u (12,000 lbs). On this basis the yield strength of this specimen is

$$(S_y)_w = L_x / A'_z = 10,000 / 0.101 = 99,200 \text{ psi}$$

And the tensile strength of this previously deformed specimen is

$$(S_u)_w = L_x / A'_z = 10,000 / 0.101 = 99,200 \text{ psi}$$

Thus *the tensile strength of a material is equal to its yield strength when it is cold-worked to a tensile prestrain that is greater than the material's original ultimate load strain.* And the yield strength of the cold-worked material can be calculated from Eq. (7-41). The above relationship can be expressed mathematically as

214 Chapter 7

$$(S_u)_w \approx (S_y)_w \approx \sigma_o(\varepsilon_w)^m \quad \text{for} \quad \varepsilon_w \geq \varepsilon_u \quad (7\text{-}43)$$

By means of Eqs. (7-42) and (7-43), the tensile strength of any material that is given any amount of tensile prestrain can be easily calculated.

Tensile Strength-to-Brinell Hardness Ratio

It is commonly known by mechanical design engineers that the tensile strength of a steel can be estimated by multiplying its Brinell hardness number by 500. This has helped lead to the wide acceptance of the Brinell hardness scale. However, this ratio is not 500 for all materials — it varies from as low as 450 to as high as 1000 for some of the commonly used metals. The ratio of the tensile strength of a material to its Brinell hardness number is identified in this book by the symbol K_b, and it is a function of both the load used to determine the hardness and the strain-strengthening exponent of the material.

Since the Brinell hardness number of a given material is not a constant (see the section "Brinell Hardness," p. 180) but varies in proportion to the applied load, it follows that the proportionality coefficient K_b is not a constant for a given material, for it too varies in proportion to the load used in determining the hardness. For example, a 50% cobalt alloy (L 605 or HS 25) has a Brinell hardness number of 200 when tested with a 3000 kg load and a hardness of only 150 when tested with a 500 kg load. Since the tensile strength is about 145,000 psi for this annealed alloy, the value for K_b is about 970 for the low load and about 730 for the high load. Likewise, a certain heat of alpha brass has Brinell hardness number of 45 when tested with a 500 kg load, but the hardness number is 50% higher, or 58 Hg, when tested with a 2000 kg load. Similar results are obtained for all metals.

Since the material is subjected to considerable plastic deformation both when the tensile strength and the Brinell hardness are measured, these two values are influenced by the strain-strengthening exponent m of the material. Therefore K_b must also be a function of m.

Figure 7-9 is a plot of experimental data obtained by the author over a number of years that shows the relationships between the ratio K_b and the variables: the strain-strengthening exponent, m, and the diameter of the indentation d, which is a function of the applied load. From these curves it is apparent the K_b varies directly with m and inversely with d. The following examples illustrate the applicability of these curves.

A test was conducted on a heat of alpha brass to see how accurately the tensile strength of a material could be predicted from a hardness test, with the strain-strengthening exponent of the material not known. Loads varying from 200 kg to 2000 kg were applied to a 10 mm ball, with the following results:

load (kg):	200	500	1000	1500	2000
diameter (mm):	2.53	3.65	4.82	5.68	6.30

When plotted on log-log paper these data fall on a straight line having a slope of 2.53, which is the Meyer strain hardening exponent p. The equation for this straight line is

$$L = 18.8 \, d^{2.53}$$

Since m = p − 2, the value of m is 0.53.

For ease in interpreting Figure 7-9, the load corresponding to an indentation of 3 mm is calculated from the above equation as 305 kg. The Brinell hardness number can be calculated from Eq. (7-1) as 43. K_b can now be determined from Figure 7-9. For m = 0.53 and d = 3 mm (d/D = 0.3), K_b is 890. Thus the tensile strength is:

$$S_u = K_b H_B = 890 \times 43 = 38,300 \text{ psi}$$

In similar fashion, the load for a 5 mm diameter is 1100 kg and the corresponding Brinell hardness number is 53. From Figure 7-9 the value of K_b is found to be 780 and the tensile strength is estimated as

$$S_u = K_b H_B = 780 \times 43 = 41,300 \text{ psi}$$

The average value of these two calculated tensile strengths is 39,800 psi. The experimentally determined value of the tensile strength for this brass is 40,500 psi, which is 2% greater than the predicted value.

As another example, consider the estimation of the tensile strength for a material when its typical strain-strengthening exponent is known. Annealed 3003 aluminum has an average m value of 0.28. What is the tensile strength of a heat that has a Brinell hardness number of 28 when measured with a 500 kg load. From Table 7-3 the diameter of the indentation for this hardness number is found to be 4.65. Then from Figure 7-9 the value of K_b is determined as 535. The tensile strength can then be calculated:

$$S_u = K_b H_B = 535 \times 28 = 15,000 \text{ psi}$$

216 Chapter 7

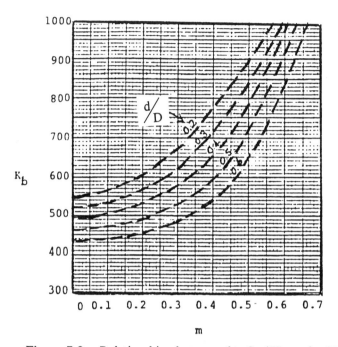

Figure 7-9. Relationships between the S_u / H_B ratio (K_b) and the strain strengthening exponent m. D is the diameter of the ball and d is the diameter of the indentation. From experimental data by the author.

If another heat had a Brinell hardness of 25 (d = 4.95 mm), it would have a K_b of 525 and its tensile strength would be 525 × 25 or 13,100 psi.

As a third example, consider the determination of the hardness of a material if its tensile strength and m value are both known. From Table B-2 (pg. 000) it is seen that annealed 17-4 PH stainless steel has a tensile strength of 142,000 psi and a strengthening exponent of 0.05. For this low value of m it can be seen in Figure 7-9 that the value of K_b must vary between about 450 and 550. This problem requires a trial-and-error solution. That is, first an estimate has to be made of the hardness and then verified. If the first estimate is wrong, then additional ones have to be made until the correct value is achieved. Some judgment must be used in making the first estimate to arrive at a value near the correct one. In this case, since the tensile strength of this PH steel is relatively high, it can be assumed that its hardness will also be relatively high. This

Table 7-3. Hardness Conversion Values

d	H_B 500	H_B 3000	R_C	R_B	H_V	H_K	d	H_B 500	H_B 3000	R_C	R_B	H_V	H_K
6	159	946					4.40	31.2	187	10	90	185	201
2.05	143	857					4.45	30.5	183		89	180	196
2.10	143	857					4.50	29.8	179		88	176	192
2.15	136	817					4.55	29.1	174		87	172	188
2.20	130	780	68		942	920	4.60	28.4	170		86	169	184
2.25	124	745	66		854	870	4.65	27.8	167		85	165	180
2.30	119	712	64		789	822	4.70	27.1	163		84	162	176
2.35	114	682	62		739	776	4.75	26.5	159		83	159	173
2.40	109	653	60		695	732	4.80	25.9	156		82	156	170
2.45	104	627	59		675	710	4.85	25.4	152		81	153	167
2.50	100	601	58		655	690	4.90	24.8	149		80	150	164
2.55	96.3	578	57		636	670	4.95	24.3	146		79	147	161
2.60	92.6	555	56		617	650	5.00	23.8	143		78	144	158
2.65	89.0	534	54		580	612	5.05	23.3	140		77	141	155
2.70	85.7	514	52		562	594	5.10	22.8	137		75	137	150
2.75	82.6	495	51		528	558	5.15	22.3	134		74	135	147
2.80	79.6	477	49		498	526							
2.85	76.8	461	48		485	510	5.20	21.8	131		73	132	145
2.90	74.1	444	47		471	495	5.25	21.4	128		71	127	141
2.95	71.5	429	46		458	479	5.30	20.9	126		70	125	139
3.00	69.1	415	44		436	452	5.40	20.1	121		68	121	135
3.10	64.6	388	42		413	423	5.45	19.7	118		67	119	133
3.15	62.5	375	40		393	402	5.50	19.3	116		65	116	131
3.20	60.5	363	39		383	391	5.55	18.9	114		64	114	129
3.25	58.6	352	38		373	380							
3.30	56.8	341	37		363	382	5.60	18.6	111		63	112	127
3.35	55.1	331	36		353	360	5.70	17.8	107		60	107	121
3.40	53.4	321	35		343	351	5.75	17.5	105		59	106	120
3.45	51.8	311	34		334	342	5.80	17.2	103		57	103	120
3.50	50.3	302	33		325	334	5.85	16.8	101		55	100	116
3.55	48.9	293	31		309	318	5.90	16.5	99		53	47	114
3.60	47.5	285	30		301	311	5.95	16.2	97		50	93	110
3.65	46.1	277	29		293	304	6.00	15.9	96				
3.70	44.9	269	28		285	297	6.05	15.6	94				
3.75	43.6	262	26		271	284	6.10	15.3	92				
3.80	42.4	255	25		264	278	6.15	15.1	90				
3.85	41.3	248	24		257	272	6.20	14.7	89				
3.90	40.2	241	22	100	240	251	6.25	14.5	87				
3.95	39.1	235	21	99	234	246	6.30	14.2	86				
4.00	38.1	229	20	98	228	241	6.35	14.0	84				
4.05	37.1	223	20	97	222	236	6.40	13.7	83				
4.10	36.2	217	18	96	216	231	6.45	13.5	81				
4.15	35.3	212	17	95	210	226	6.50	13.3	80				
4.20	34.4	207	16	94	205	221							
4.25	33.6	201	15	93	200	216							
4.30	32.8	197	13	92	195	211							
4.35	32.0	192	12	91	190	206							

Conversion from diameter of indentation to H_B is valid for all materials. Conversion from 3000 kg H_B to other hardness scales is valid only for steel,

$$H_B = 2L / \pi D \left[D - \left(D^2 - d^2 \right) \right]^{0.5}$$

; d = diam. of indentation, mm; H_B = Brinell hardness; R_C and R_B = Rockwell hardness; H_V = Vickers or Diamond hardness; H_K = Knoop hardness.

means that for a given load the diameter of the indentation will be small. On this basis the first assumption will be made that d/D = 0.3 and therefore K_b = 530. With a d of 3.00 mm, the Brinell hardness is 415 for a 3000 kg load. To check the accuracy of this first assumption, the tensile strength is calculated as 530 × 415, or 220,000 psi. This is much greater than the experimental value of 143,000 psi, so the first assumption is wrong. A brief analysis shows that this hardness is too high.

In order to obtain a lower hardness, a second assumption is made that d = 4.00 mm. This gives a K_b of 500 in Figure 7-9 and an H_B of 229 in Table 7-3. To check this assumption, the tensile strength is now calculated as 500 × 229 or 114,500 psi. This value is too low. For a third assumption, assume d = 3.60 mm. This gives K_b = 510, H_B = 285, and a tensile strength of 510 × 285 = 145,000 psi. Since this is just slightly high, the correct value of the Brinell is about 280.

Although the value of K_b for 17-4 PH is 510, which is close to the value of 500 for the common steels, it must not be concluded that all stainless steels have K_b values in this range. Since the strain-strengthening exponents of the various grades of austenitic stainless steel vary between 0.4 and 0.5, their K_b values are in the range of 700 to 800.

As a fourth example, consider the plain carbon steels. Why is it possible for K_b to be 500 for all of the ordinary steels even though their strain-strengthening exponents are not the same? The value of m varies from about 0.27 for annealed 1020 steel to about 0.08 for annealed 1080 steel, yet K_b is 500 for both of them. The answer lies in the fact that the 1080 steel is harder and therefore has a smaller d/D ratio. To verify this, it is simply necessary to check out the appropriate values in Figure 7-9. The average value of H_B for 1020 steel is 120. This corresponds to a 5.40 mm diameter indentation. For an m value of 0.27 and d/D of 0.54, the value for K_b is found to be 500 in Figure 7-9.

The average value of H_B for 1080 steel is 240, which corresponds to a 3.92 mm indentation. The value of K_b is again found to be 500 in Figure 7-9 for an m value of 0.08 and d/D = 0.392.

The Effect of Plastic Deformation on σ_o and m

The conventional analysis. *(σ_o and m vary with cold work).* The current practice or conventional approach to explaining the effect that cold work has on the numerical values of σ_o and m is based on the work of early researchers such as Mehringer and MacGregor[2]. In this test they conducted true stress-strain tensile testes on specimens cut from an annealed low carbon steel plate. From this same plate they cut out a number of pieces that were cold rolled to reductions of 10, 20, 30, 40, etc. percent. Tensile specimens were then machined from these cold rolled pieces and subjected to the tensile tests. The true stress-strain data was plotted on logarithmic coordinates with each specimen treated as a virgin specimen.

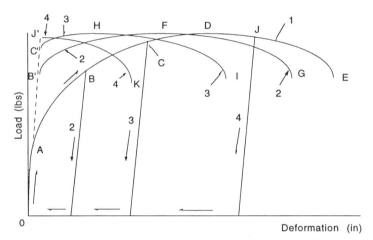

Figure 7-10. Load vs. deformation. Interrupted tensile test in which the previous tensile strain is ignored and each specimen is retested as a virgin specimen.

From this study it was found that all of the specimens followed the exponential relationship $\sigma = \sigma_0 \varepsilon^m$. Furthermore, the value of σ_0 increased slightly with an increase in the amount of cold work, and the value of m decreased significantly as the amount of cold work increased.

To explain this phenomenon, the author conducted the following demonstration. Four tensile specimens from the same bar of annealed metal were machined. The first specimen was subjected to a standard monotonic tensile test. It was elastic from O, the origin, to point A, where it yielded, as shown in Figure 7-10. The load continued to increase to point D, the ultimate

---

[2]F.J. Mehringer and C.W. MacGregor, Effects of cold-rolling on the true stress-strain properties of low-carbon steel. <u>Journal of AIME</u>, Feb., 1945 p 291.

load, and then decreased until the specimen fractured at E. This behavior is illustrated as curve 1. Simultaneous readings of load and deformation were recorded during this and the subsequent tests.

The second specimen was subjected to an interrupted tensile test. It was loaded along the path OAB, which coincides with the path of the first specimen. Upon reaching load B, it was immediately unloaded to a zero load. Following the unloading, it was retested as a virgin specimen and followed the path OAB' where the new yield load was B' which was numerically equal to the previous highest load B. As the load was increased, the second specimen reached a maximum load at F, which was numerically equal to the ultimate load D of the first specimen, but at a lesser deformation. The load then dropped and the specimen fractured at G, which was numerically equal to the load of E, but also with a lesser deformation.

An important observation in regard to the deformation of the second specimen is that the sums of the original deformation to reach point B plus the additional deformation required to reach points F and G are equal to the deformation of the points D and E, respectively, of the first specimen.

In like fashion, the third specimen was loaded to point C, unloaded, and retested. The new yield load was at C', equal to C, and the highest load was at H, equal to D. The fracture load was at I, equal to E.

The fourth specimen was loaded beyond the ultimate load D, where necking of the specimen began, to point J, and immediately unloaded. Upon reloading, the specimen now reached both its yield load and maximum load at the same value of J' equal to J.

When the above data, shown as load vs. deformation in Figure 7-10, are reduced to true stress and true strain and are plotted on log-log coordinates, the curves shown in Figure 7-11 are obtained.

The data from the first specimen give the elastic line OA, and the plastic behavior is the line AE. The equation for this line is $\sigma = \sigma_0 \varepsilon^m$.

The plastic region for the second specimen is B'G and fits the equation $\sigma = (\sigma_0)_2 \varepsilon^{m_2}$. The plastic region for the third specimen is C'I and fits the equation $\sigma = (\sigma_0)_3 \varepsilon^{m_3}$. The line for the fourth specimen is J'K and fits the equation $\sigma = (\sigma_0)_4 \varepsilon^{m_4}$.

If the fracture strain associated with the first specimen at point E is designated as ε_f, then the value for the fracture strain at point G for the second specimen is equal to $\varepsilon_f - \varepsilon_B$, where ε_B is the value of prestrain given the specimen at point B. Likewise, the fracture strain for the third and fourth specimen is equal to $\varepsilon_f - \varepsilon_C$ and $\varepsilon_f - \varepsilon_J$, respectively, where ε_C is the prestrain given the third specimen and ε_J is the prestrain of the fourth specimen.

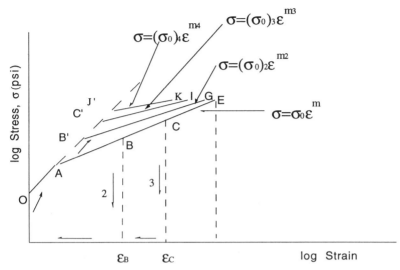

Figure 7-11. True stress vs. true strain for the tensile tests shown in Figure 7-10.

If the strain at the ultimate load D for the first specimen is ε_u, then the value for the ultimate load strain at point F for the second specimen is equal to $\varepsilon_u - \varepsilon_B$. Likewise, the value for the ultimate load strain at point H for the third specimen is equal to $\varepsilon_u - \varepsilon_C$, and the ultimate load strain of the fourth specimen is zero.

There is a significant disadvantage to this method. The material must be subjected to varying degrees of cold work, and tensile specimens must be machined from each of the cold-worked materials. And finally, the tensile tests must be conducted and the data reduced to obtain the values for $(\sigma_0)_1$, $(\sigma_0)_2$, $(\sigma_0)_3$, m_1, m_2, m_3, etc.

The Datsko-Mitchell analysis. *(σ_0 and m do not change with cold work).* This analysis is based on the results of interrupted tensile tests and is valid for straining in the same sense and direction as that of the prestrain. In this method the prestrains $\varepsilon_B, \varepsilon_C$, and ε_J are included with the subsequent tensile strains, as illustrated in Figures 7-12 and 7-13. In other words, specimens 2, 3, and 4 are

222   Chapter 7

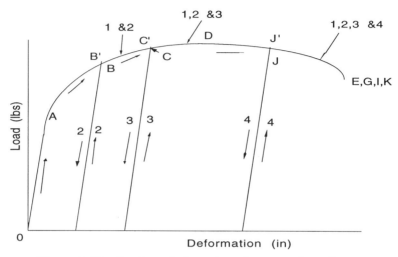

Figure 7-12. Load vs. deformation. Interrupted tensile test where the previous tensile strain is included and each specimen is retested as a continuation of the original test.

retested as a continuation of their previous loading. The curves shown here represent the actual results obtained in the laboratory.

It is evident in Figure 7-13 that the numerical values of σ_0 and m do not change when the value of the tensile prestrain ε_w is included in the relationship as $\sigma = \sigma_0(\varepsilon + \varepsilon_w)^m$. The advantage of this method is that it eliminates the expense and effort of cold working the metal, machining additional tensile specimens, and subjecting them to tensile testing and analysis.

However, engineering analysis sometimes requires knowing what the numerical values of $(\sigma_0)_w$ and m_w would be for a given amount of plastic deformation. Two such instances are (1) prediction of the effect of cold work on the fatigue strength of a metal, and (2) prediction of the tensile strength to Brinell hardness ratio of a metal.

Predicting σ_0 and m After Tensile Deformation

The author has developed the following method to predict the values of $(\sigma_0)_w$ and m_w from the original non-cold-worked tensile properties of σ_0 and m and

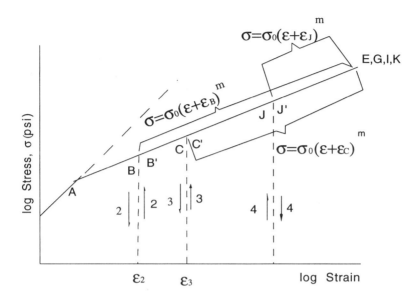

Figure 7-13. True stress vs. true strain plots of the tensile tests illustrated in Figure 7-12.

the amount of cold work ε_w. This method avoids the need for experimental testing. The model is described in reference to Figure 7-14.

Line OA represents the original elastic properties of the metal, $\sigma = E\varepsilon$. Line AF represents the original plastic properties, $\sigma = \sigma_o \varepsilon^m$. ε_f and σ_f are the fracture strain and fracture stress, respectively, of the original metal.

A second identical specimen is loaded to B, where the stress is σ_B and the strain is ε_B. It is unloaded and reloaded along OAB′. The new yield strength, $(S_y)_w$, is equal to σ_B but it occurs at a strain that is equal to the sum of the elastic strain, σ_B/E, plus the offset strain of 0.002 in./in. Thus point C is 0.002 in./in. to the right of point B′. The coordinates of point C are $\sigma_C = \sigma_B$ and $\varepsilon_c = (\sigma_B / E) + 0.002$.

On continued loading to fracture, the metal deformation follows the line CD. Since the fracture stress σ_f of a material is a physical property of the material, $(\sigma_f)_w = \sigma_f$. That is, the fracture stress of the load-interrupted specimen is the same as the fracture stress of the original specimen.

224 Chapter 7

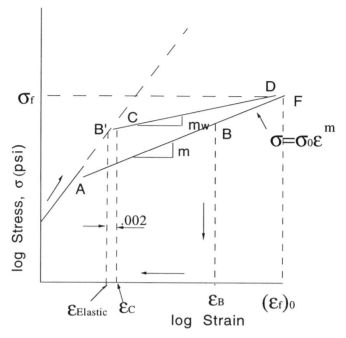

Figure 7-14. Illustration of the method to calculate m_w from original non-cold-worked tensile properties.

Also, since true strains can be added or subtracted if they are in the same direction, the fracture strain of the reloaded specimen $(\varepsilon_f)_w$ is equal to $(\varepsilon_f)_0 - \varepsilon_B$. Thus the coordinates of point D are $\sigma_D = \sigma_f$ and $(\varepsilon_f)_D = (\varepsilon_f)_0 - \varepsilon_B$.

The equation for the line CD is $\sigma = (\sigma_0)_w \varepsilon^{m_w}$. By substituting the coordinates of point C, the equation is $\sigma_C = (\sigma_0)_w \varepsilon_C^{m_w}$, and for point D the equation is $\sigma_D = (\sigma_0)_w \varepsilon_D^{m_w}$. By solving the simultaneous equations the value for m_w is found to be equal to $\ln(\sigma_C / \sigma_D) / \ln(\varepsilon_C / \varepsilon_D)$. In terms of the original properties of the metal, the value of m_w is:

$$m_w = \frac{\ln\left[\sigma_o(\varepsilon_B)^m / \sigma_o(\varepsilon_f)^m\right]}{\ln\left[\left(\sigma_o(\varepsilon_B)^m / E + 0.002\right) / (\varepsilon_f - \varepsilon_B)\right]} \quad (7\text{-}44)$$

The value for $(\sigma_o)_w$ can then be calculated.

FATIGUE PROPERTIES OF METALS

The previously discussed tensile properties are frequently referred to as static properties because only one slowly applied load is placed on the material during the test. But many structural components are subjected to dynamic loads, or loads that are reapplied a large number of times. If a load is applied to a part many times, rather than just once, the part may fail even though the load on it is less than the ultimate load (that is, the stress on it is less than its tensile strength). Many materials will fail with a load that is as low as one-half the ultimate load if it is applied a million times or more. The load may vary from zero to some tensile value, or from tension to compression. A failure that results from such cyclic loads is called a fatigue failure. Since many structural components are subjected to cyclic loads it is necessary for the design engineer to have some quantitative measure of the material's ability to withstand such repeated loads.

Quantitative data for the fatigue properties of a given material are obtained by subjecting a number of standard specimens to cyclic loads until fracture occurs. One specimen (preferably two or three to observe the amount of scatter in the data) is tested at each of several stress levels varying between approximately $0.9 S_u$ and $0.5 S_u$, and the number of cycles of load that are required to cause fracture is recorded. Table 7-4 illustrates such data for three different steels.

Fatigue strength data is usually presented in graphical rather than tabular form. The data in Table 7-4 are plotted on semi-log coordinates in Figure 7-15. The data may also be plotted on log-log coordinates. The following definitions regarding the fatigue properties of materials can best be understood in conjunction with Figure 7-15.

226 Chapter 7

Table 7-4 Fatigue Strength Data for Steel

Number of cycles to failure (N)	Applied Stress (ksi)		
	Alloy steel (Q&T)	Carbon steel (Q&T)	1020 steel (HR)
10,000	108	71	45
100,000	90	58	39
1,000,000	73	44	33
10,000,000	70	44	32
100,000,000	70	44	32

Fatigue Strength

The *fatigue strength* is the value of the alternating stress that results in failure by fracture after a specific number of cycles of load application. It can also be the ordinate of the $\sigma - N$ (stress vs. number of cycles to failure) curve. Most machine design textbooks state that the fatigue strength of most metals at 10^3 cycles is approximately 0.9 times the tensile strength.

Endurance Limit

The *fatigue limit*, or *endurance limit,* is the largest value of alternating stress that will not result in fracture, regardless of the number of cycles of applied load. It is the value of alternating stress corresponding to the horizontal portion of the $\sigma - N$ curve. In this case the endurance limit is defined as the alternating stress that causes failure after some specified number of cycles. For example, the aluminum alloys do not have a horizontal portion to their $\sigma - N$ curve. Consequently, the Aluminum Association specifies the endurance limit of the aluminum alloys as that value of alternating stress that causes fracture in a rotating beam specimen at 5×10^8 cycles of load. The fatigue limit of most steels is approximately one-half the tensile strength.

Fatigue Ratio

The *fatigue ratio* is the ratio of the fatigue strength to the tensile ultimate strength, S_e / S_u.

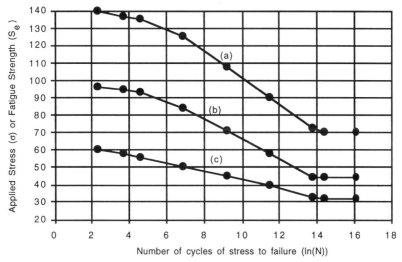

Figure 7-15. Fatigue data for three steels. (a) Alloy steel, quenched and tempered; (b) carbon steel, quenched and tempered; (c) 1020 HR steel.

Three different types of specimens are commonly used to evaluate the fatigue characteristics of a material. One type is referred to as the R. R. Moore rotating-beam specimen. It somewhat resembles a round tensile specimen whose gage section (in a plan view) is not a straight line parallel to its axis, but rather it an arc of a circle of 9 7/8" radius, with the smallest diameter at the mid-length of the specimen. This specimen is subjected to a load that causes bending, with the top fibers of the specimen in compression and the bottom fibers in tension. The magnitude of the bending stress varies from a maximum at the outermost fibers to zero at the axial centerline. As the specimen rotates during the test, each fiber at the surface is alternately subjected to tensile and compressive stresses.

The second type of specimen is a cantilever-beam type. It is similar to the rotating beam specimen in that the outer fibers are alternately subjected to tensile and compressive stress and the magnitude of the stresses varies from zero at the mid-thickness to a maximum at the outer fibers. But, instead of rotating, one end of the beam is fixed and the other end is forced to move above and below its equilibrium position. Cantilever-beam specimens are flat, having a width that is many times the thickness. There are two types of cantilever-beam specimens. One has a tapered width in the gage section so that the stress is constant throughout the gage length. This type is referred to as a constant-stress fatigue specimen. The fatigue crack can originate on the surface at any axial distance in the gage section. The second type of cantilever-beam specimen has a reduced width at one location along its length made by machining two circular sections, one from each edge of the specimen. The fatigue fracture always occurs at this reduced section.

The third type of fatigue specimen is loaded uniaxially, and therefore it has a uniform stress across the entire cross section. This type of specimen resembles a short tensile specimen. The values of fatigue strength for a given material obtained from these three types of specimens are not the same; therefore, it is necessary in designing a part against fatigue failure to use data obtained from the type of specimen that most closely resembles the type of loading that a specific part is subjected to.

Several types of load cycles are employed to determine the fatigue strength of materials. The three most common load cycles are illustrated in Figure 7-16, including a fluctuating tensile load that results in tensile stresses where the minimum value is greater than zero (a), a minimum tensile stress that is equal to zero (b), and a reversed load cycle where the material is alternately subjected to tensile and compressive stresses (c). Many modifications of these three load cycles can be employed, including a randomly varied load of varying intensity.

Intrinsic Fatigue Strength

The *intrinsic fatigue strength* of a material is obtained when the test results are not influenced by the condition of the specimen or the testing apparatus. Thus, the intrinsic fatigue strength is the true fatigue strength of a material. Unfortunately most fatigue tests simply determine a material's fatigue strength and not its intrinsic fatigue strength. This is because most fatigue data are collected in such a manner that the test conditions influence the results. The following are the most common factors that affect the test results this way.

The surface properties of the specimen should not be altered during the fabrication of the specimen. Machining, for example, leaves a thin surface layer of material that is severely cold worked and has large residual stresses. Because the process of machining a fatigue specimen has an adverse effect on its strength, if the intrinsic fatigue strength of annealed 1020 steel is to be determined, it is necessary to anneal the fatigue specimen after it is machined rather than to machine a specimen from the annealed steel.

The piece of material that the specimens are made of should be typical of the class of material that is being evaluated. Conditions such as decarburization and impurity inclusions must be specified and controlled.

The environment in which the fatigue test is conducted must be the same as that in which the actual part operates. Fatigue data obtained in room

Mechanical Properties 229

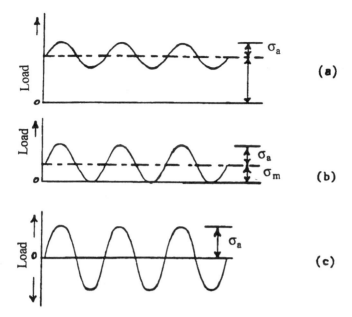

Figure 7-16. Types of fatigue load cycles. (a) Fluctuating tension load cycle, $\sigma_m > \sigma_a$. (b) Fluctuating tension load cycle, $\sigma_m = \sigma_a$. (c) Reversed load cycle, $\sigma_m = 0$.

temperature air is not suitable for the design of a part that operates in a salt spray atmosphere, at a different temperature.

The surface of the specimen must be free of notches, scratches, or deep machining marks. These serve as stress raisers and give erratic test results.

There are several theories regarding the mechanism by which fatigue failures occur. Some of them are based on the concept that small fatigue cracks form in the material either at inclusions, voids, or notches, which serve as stress raisers. Some are based on the concept that the repeated cyclic stressing induces the accumulation of small slip movements into intrusions (depressions or notches) or extrusions (projections). These small fatigue cracks usually originate at the surface of the material and slowly grow across the section. Eventually the crack becomes so large that the stresses across the uncracked section approach the tensile strength of the material, and then complete separation of the part occurs instantly. Very little plastic deformation occurs in the vicinity of the original fatigue crack, but sometimes considerable deformation, even necking, occurs during the final fracture.

IMPACT STRENGTH

In some cases a structural part is subjected to a single large, suddenly applied load. A standard test has been devised to evaluate a material's ability to absorb the impact energy through plastic deformation. The test can be described as technological (or applied) like the Rockwell hardness, rather than scientific. The values obtained by the impact test are relative rather than absolute and serve as a basis of comparison and specification of a material's toughness.

The *impact strength* is the energy, expressed in foot-pounds, required to fracture a standard specimen with a single impact blow. The impact strength of a material is frequently referred to as being a measure of a material's toughness; that is, its ability to absorb energy. The area under the tensile stress-strain curve is also a measure of a material's ability to absorb energy (its toughness). Unfortunately, there is only a very general relationship between these two different measures of toughness; namely, if a material has a large area under its tensile stress-strain curve, it also has a relatively high impact strength.

Most impact strength data are obtained with the two types of notched specimens shown in Figure 7-17: (a) illustrates the Charpy V-notch specimen as well as how the impact load is applied; (b) does the same for the Izod V-notch specimen; (c) shows the details of the notch. There are several modifications of the standard V-notch specimen. One is called the keyhole notch and another the U-notch. Both of these have a 1 mm radius at the bottom rather than the 0.25 mm radius of the V-notch. There is no correlation among the various types of notch-bar impact strength values. However, the Charpy V-notch impact strength is considerably greater than the Izod V-notch value, particularly in the high toughness range.

The impact-testing machine consists of a special base mounted on the floor to support the specimen, and a striking hammer that swings through an arc of about a 32" radius, much like a pendulum. When the hammer is "cocked" (raised to a locked elevation) it has a potential energy that varies between 25 and 250 ft lb, depending upon the mass of the hammer and the height to which it is raised. When the hammer is released and allowed to strike the specimen, a dial registers the energy that was absorbed by the specimen. The standards specify that the striking velocity must be in the range of 10 to 20 ft per sec because velocities outside this range have an effect on the impact strength.

The impact strength of some metals, particularly steel, varies significantly with the testing temperature. Figure 7-18 shows this variation for a normalized AISI 1030 steel. At the low testing temperature, the fracture is of the cleavage type, which has a bright, faceted appearance. At the higher temperatures the fractures are of the shear type, which have a fibrous appearance. The transition temperature is that temperature that results in 50% cleavage fracture and 50% shear fracture. It may also be defined as the temperature at which the impact

Mechanical Properties   231

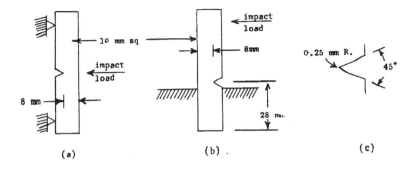

Figure 7-17. Impact tests and specimens: (a) Charpy L = 55 mm; (b) Izod L = 75 mm; (c) details of the notch.

strength shows a marked drop. The nil ductility temperature is the highest temperature at which the impact strength starts to increase above its minimum value. These two temperatures are illustrated in Figure 7-18.

CREEP STRENGTH

A part may fail with a load that induces stresses that lie between the yield strength and the tensile strength of the material even if the load is steady and constant, rather than alternating and repeating as in a fatigue failure. This type of loading causes the part to slowly elongate, or creep. The failure point may be when the part stretches to some specified length or when the part completely fractures.

The *creep strength* of a material is the value of nominal stress that will result in a specified amount of elongation at a specific temperature in a given period of time. It is also defined as the value of nominal stress that induces a specified creep rate at a specific temperature. The creep strength is sometimes called the *creep limit*. The *creep rate* (Figure 7-19) is the slope of the strain-time creep curve in the steady creep region referred to as stage 2 creep.

Most creep failures occur in parts that are exposed to high temperatures. The stress necessary to cause creep at room temperature is considerably higher than the material's yield strength. In fact, it is just slightly less than the material's tensile strength. The stress necessary to induce creep at a temperature that is higher than the material's recrystallization temperature, on the other hand, is very low.

The specimens used for creep testing are quite similar to round tensile specimens. During the creep test, the specimen is loaded with a dead weight

232   Chapter 7

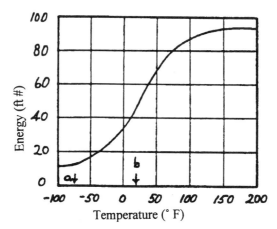

Figure 7-18. Charpy V-notch impact strength of 1030 steel vs. temperature: (a) nil ductility, (b) transition temperature.

that induces the required nominal stress that is applied throughout the entire test. The specimen is enclosed in a small tube-type furnace to maintain a constant temperature throughout the test, and the gage length is measured after various time intervals. The three variables that affect the creep rate of the specimen are (1) the nominal stress, (2) the temperature, and (3) the time.

Figure 7-19 illustrates the most common method of presenting creep test data. Three different curves are shown. Curve (a) is typical of a creep test conducted at a temperature well below the material's recrystallization temperature (room temperature for steel) and at a fairly high stress level, slightly above the material's yield strength. Curve (a) is also typical of a creep test conducted at a temperature near the material's recrystallization temperature but at a low stress level. Curve (c) is typical of either a high stress (slightly below the material's S_u) at a low temperature or else a low stress level at a temperature way above the material's recrystallization temperature. Curve (b) illustrates the creep rate at an intermediate combination of stress temperature.

A creep curve consists of four separate parts as illustrated by curve (b) of Figure 7-19 and explained below:

1. An initial elastic extension from the origin O to point b.
2. A region of primary creep, frequently referred to as stage 1 creep. The extension occurs at a decreasing rate in this portion of the creep curve.

Mechanical Properties 233

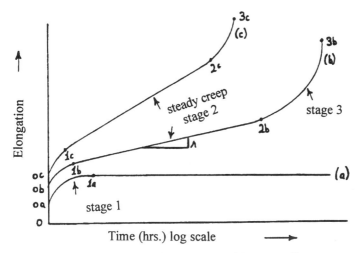

Figure 7-19. Creep data plotted on semi-log coordinates. (a) Low stress (slightly above S_y) or low temperature (well below recrystallization). (b) Moderate stress (midway between S_y and S_u) or moderate temperature (at recrystallization). (c) High stress (slightly below S_u) or high temperature (well above recrystallization). oa, ob, and oc are elastic elongation.

3. A region of secondary creep, frequently called stage 2 creep. The extension occurs at a constant rate in this region. Most creep design is based on this portion of the creep curve since the creep rate is constant and the total extension for a given number of hours of service can be easily calculated.
4. A region of tertiary creep, or stage 3 creep. In this region, the extension occurs at an increasing rate until the material fractures.

Another practical way of presenting creep data are illustrated in Figure 7-20, which is a log-log plot of nominal stress versus the second stage creep rate, expressed as percent per hour, with the temperature as a parameter. Figure 7-21 illustrates still another type of plot used to present creep data in which both the stress and temperature are drawn on cartesian coordinates.

The mechanism of creep is very complex inasmuch as it involves the movements of vacancies and dislocations, strain hardening, and recrystallization, as well as grain boundary movements. At low temperatures creep is restricted by the pileup of dislocations at the grain boundaries and the resulting

234 Chapter 7

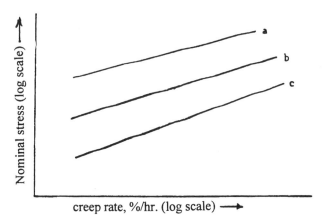

Figure 7-20. Second stage creep rate vs. nominal stress. a, b, c are for low, medium, and high temperatures, respectively.

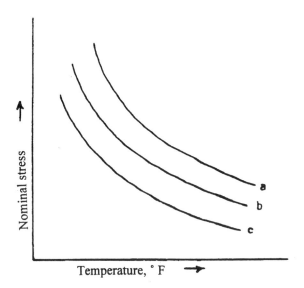

Figure 7-21. Second stage creep rate vs. temperature and nominal stress. a = 1 %/hr creep rate; b = 0.1 %/hr creep rate; c = 0.001%/hr creep rate.

Mechanical Properties 235

strain hardening. But at the higher temperatures the dislocations can climb out of the original slip plane and thus permit further creep. In addition, recrystallization with its resulting lower strength permits creep to occur readily at higher temperatures.

As explained in Chapter 2, the grain boundary material is stronger than the material of the interior portions at low temperatures, but the opposite is true at high temperatures. The temperature at which the strength of grain boundaries and interior grains is equal is called the equicohesive temperature. Consequently, parts that are exposed to high temperatures have lower creep rates if they are originally heat treated to form coarse grains.

ESTIMATING LOW CYCLE FATIGUE STRENGTH FROM TENSILE TEST DATA

The Current Method

One of the most common methods of estimating low-cycle fatigue strength of metals today is to construct a curve on semi-log paper with the ordinate as strength and the abscissa as number of cycles to failure. This is illustrated in Figure 7-22 for a steel having a tensile strength of 60 ksi and an endurance limit of 30 ksi.

The tensile strength is plotted on the 1 cycle line as point 1. The endurance limit is plotted as 30 ksi on the 10^6 cycles line as point 3. Point 2 is estimated by assuming failure at 10^3 cycles at a stress level that is 0.9 times the tensile strength (sometimes the value 0.85 is used). Thus, point 2 is plotted at 54 ksi and on the 10^3 cycles line. Next, straight lines are drawn between points 1 and 2 and points 2 and 3. A horizontal line is extended from point 3. This resulting curve is then used to predict the fatigue strength at any intermediate number of cycles.

Datsko's Method

The author has derived another method of predicting fatigue strength, or the stress that will cause failure under cyclic loading, that appears to be more meaningful and reliable than the preceding method. It is based on stress rather than strength and utilizes the plastic properties of the material as determined from a tensile test, namely; the strength coefficient σ_0 the strain strengthening exponent m, and the true strain at fracture ε_f. The plastic stress strain characteristics are best portrayed by the expression $\sigma = \sigma_0 \varepsilon^m$, where σ is the

236   Chapter 7

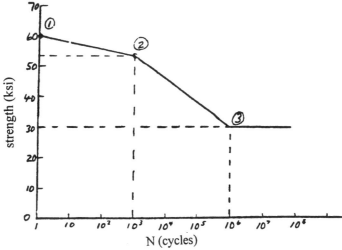

Figure 7-22. Estimation of low-cycle fatigue strength for steel having a tensile strength of 60 ksi and an endurance limit of 30 ksi (50% of S_u).

stress and ϵ is the natural or logarithmic strain defined as the natural logarithm of the ratio of instantaneous to original lengths. That is:

$$\epsilon = \log_e(l_i / l_o) \quad \text{or} \quad \epsilon = \ln(l_i / l_o)$$

This new method is based on the failure theory[3] developed by the author that states that "the maximum deformation that a material in a given condition can be subjected to in a plastic deformation process is that deformation that results in a natural tensile strain being induced in some direction in the part which is equal to the natural strain at fracture of a tensile specimen of that particular condition of the material." This theory establishes one point on a stress-cycles-to-failure curve: it is the stress that causes failure in the least possible number of cycles. The least number of cycles that can be applied is one-half: that is, one application of a load, which is a quarter cycle, plus another quarter cycle resulting from the unloading. This is based on the concept of

---

[3]Datsko, J. Material Properties and Manufacturing Processes. John Wiley, 1965, p 315.

Mechanical Properties   237

reversed cyclic loading, where the second half cycle comes when the opposite (compressive) load is applied.

This concept is applied to bending fatigue specimens in the following manner. When a specimen is bent, the outer fibers are strained in tension. When the outer fiber strain is numerically equal to ε_f for the material, the specimen will fail with just the half cycle of load. The stress on the specimen at the time of failure (at the outer fiber) is equal to the true fracture stress, σ_f, which is determined from the material's tensile properties according to the relationship

$$\sigma_f = \sigma_o (\varepsilon_f)^m \tag{7-45}$$

This establishes one point on a $\sigma - N$ curve, shown as point 1 in Figure 7-23. The specific numerical values are calculated as follows. The steel of Figure 7-22 is typical of annealed AISI 1020. The typical plastic tensile properties are:

$$S_u = 60 \text{ ksi}$$
$$\sigma_o = 110 \text{ ksi}$$
$$m = 0.25$$
$$\varepsilon_f = 1.1$$

Therefore, according to Eq. (7-45), $\sigma_f = 110(1.11)^{0.25} = 113 \text{ ksi}$. Thus, point 1 is plotted at $\sigma = 113$ and $N = 0.5$.

The $\sigma - N$ curve must start at point 1 and move to the right and downward. Several shapes for this curve can be assumed, or else experimental data can be plotted, in which case the shape is empirically determined. The author tried this approach for annealed pure aluminum for a research project in which the effect of cold work on fatigue strength was being determined. A good fit of the experimental data was obtained when a straight line was drawn through the data points, which were plotted on log-log paper. The equation of this line, on the basis of the material's plastic tensile properties, is:

$$\sigma_a = \sigma_o (\varepsilon_f)^m (2N)^{-s} \tag{7-46}$$

This relationship is simply the equation of a line through the point $\sigma = \sigma_f$ and $N = 1/2$ with a negative slope s. Since the experimental data on soft aluminum resulted in a straight line, the author decided to see what kind of curves would result if published values of fatigue data were plotted in similar fashion. Several well-documented sources gave $\sigma - N$ data for annealed 1040 steel and 1045 steel heat treated to a range of hardnesses. The published data points fell on a line that went through $\sigma = \sigma_f$ at $N = 1/2$. The author also found some data for 17-4 PH stainless steel, 2024 T4 aluminum, pure copper, and 70 Cu-30 Zn brass. When all of these materials were plotted on the same set of coordinates, two facts appeared obvious: 1, The height of the curves was proportional to σ_o, the strength coefficient; 2, The slope of curves was proportional to m, the strain strengthening exponent. Both of these conditions appear to be very reasonable.

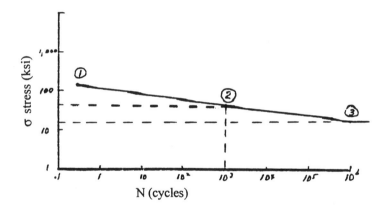

Figure 7-23. Method of predicting low cycle fatigue strength from tensile test data.

From a detailed study of the above relationships the author found that the slope s of the $\sigma - N$ curve could be determined from the tensile test data by the expression

$$s = 0.14 m^{0.25} \qquad (7\text{-}47)$$

Thus the semi-analytical Eq. (7-46) can be rewritten as

$$\sigma_a = \sigma_o(\varepsilon_f)^m (2N)^{-0.14m^{0.25}} \qquad (7\text{-}48)$$

where σ_a is the alternating stress that will result in failure in N cycles.

By means of Eq. (7-48), the alternating stress that will cause failure in 1000 cycles can be predicted as follows:

$$\sigma_{10^3} = 110(1.11)^{0.25}(2000)^{-0.14 \times 0.25^{0.25}} = 54 \text{ ksi}$$

It is a mere coincidence that this value is exactly equal to $0.9\,S_u$. For materials with very low m values, σ_{10^3} is less than $0.9\,S_u$; and for materials with very high m values, σ_{10^3} is greater than $0.9\,S_u$ and, in fact, may be greater than S_u.

In similar fashion, the endurance limit or alternating stress that will cause failure in a million cycles is determined to be:

$$\sigma_{10^6} = 110(1.11)^{0.25}(2 \times 10^6)^{-0.14 \times 0.25^{0.25}} = 27 \text{ ksi}$$

This value is in the range of published values for annealed 1020 steel.

σ_a for numbers of cycles greater than 10^6 is equal to the 10^6 value for steels since the $\sigma - N$ curve is horizontal in this region.

The big difference between the old method of estimating low-cycle fatigue strengths and the new method presented here is in the values obtained for the very low number of cycles, such as 10. Figure 7-22 indicates failure will occur at an alternating stress of about 57 ksi whereas Eq. (7-48) and Figure 7-23 indicate that the stress would have to be nearly 100 ksi. Experimental data indicate that the latter value is more correct.

Another advantage of this new method of predicting low cycle fatigue strength is that it also enables one to estimate the high-cycle fatigue strength when experimental data are lacking.

A final illustration of the usefulness of Eq. (7-48) is in comparing the σ_a / S_u ratio for materials having very different m values. Consider an 18-8 annealed stainless steel having $\sigma_o = 230$ ksi, $m = 0.53$ and $\varepsilon_f = 1.38$. By means of Eq. (7-48), $\sigma_{10^3} = 110$ ksi.

By means of the equation $S_u = \sigma_o(m/e)^m$ the value for S_u is found to be 97.6 ksi.

240   Chapter 7

Therefore, $\sigma_{10^3}/S_u = 1.14$. This means a stress that is 14% greater than the tensile strength is required to cause 18-8 stainless steel to fail in 1000 cycles.

For this same material, $\sigma_{10^6} = 48.6$ ksi and $\sigma_{10^6}/S_u = 0.50$.

For a 17-4 PH stainless steel aged at 1100°F, the tensile properties are: $\sigma_0 = 260$ ksi, $m = 0.01$, $\varepsilon_f = 0.65$, and $S_u = 246$ ksi. These values give $\sigma_{10^3} = 185$ ksi and $\sigma_{10^6} = 137$ ksi. The resulting ratios $\sigma_{10^3}/S_u = 0.75$ and $\sigma_{10^6}/S_u = 0.56$.

EXAMPLE PROBLEMS

7-1. A particular metal in the annealed condition possesses the following characteristics:

$$E = 10 \times 10^6 \text{ psi}; \quad S_y = 18{,}000 \text{ psi}; \quad S_u = 30{,}000 \text{ psi}$$
$$A_r = 40\%; \quad \sigma = 60{,}000\varepsilon^{0.33}$$

a. When subjected to a tensile strain-hardening treatment designated as X, the resultant yield strength is 30,000 psi. Determine the amount of cold work imparted by the X treatment.
b. If a tensile specimen were machined from a bar of this material that was subjected to treatment X and was tested on a tensile machine, what would the true tensile strain at fracture most likely be?
c. What will be the probable value of the percent reduction of area of the specimen in b?
d. What will be the probable value of the true fracture strength of the specimen in b?

Mechanical Properties 241

Solution:

a. $S_y = \sigma_y = 30 = 60\varepsilon_x^{0.33}$; $\varepsilon_x = 0.5^3 = 0.125$
   $\varepsilon_x = \ln(100/100-W) = 0.125$; $W = 11.75\%$
   or if $\varepsilon_x = \varepsilon_W + 0.002$ (0.2% offset), then W=11.6%

b. Original $\varepsilon_f = \ln(100/60) = 0.511$. Since natural strains are additive, the remaining strain $(\varepsilon_f)r = 0.511 - 0.125 = 0.386$.

c. $0.386 = \ln(100/100 - A'_r)$; $A'_r = 32\%$

d. $\sigma_f = 60 \times 0.511^{0.33} = 48$ ksi

7-2. 1020 annealed steel consisting of 3/4 α and 1/4 Pc is subjected to 20% cold work by rolling. Estimate its 3000 kg load Brinell hardness. For annealed ferrite: $\sigma_o = 77$, $\varepsilon_u = 0.28$, $\varepsilon_f = 1.6$.
For annealed pearlite: $\sigma_o = 172$, $\varepsilon_u = 0.11$, $\varepsilon_f = 0.5$.

Solution:
In order to use Figure 7-9, it is necessary to know the value of m_w, the strain-strengthening exponent of the cold worked steel. For 1020 annealed steel, based on microstructure:

$\sigma_o = 3/4 \times 77 + 1/4 \times 172 = 101$; $\varepsilon_u = 3/4 \times .28 + 1/4 \times 0.11 = 0.24$;
$\varepsilon_f = 3/4 \times 1.60 + 1/4 \times 0.5 = 1.33$

point 1:
$\sigma_1 = \sigma_w = 101(0.223)^{0.24} = 70.5$ $\varepsilon_1 = \dfrac{70.5}{30 \times 10^3} + 0.002 = 0.0053$

point 2:
$\sigma_2 = \sigma_f = 101(1.33)^{0.24} = 108$ $\varepsilon_2 = 1.33 - 0.223 = 1.107$

242   Chapter 7

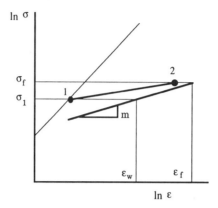

$108 = (\sigma_o)_1 (1.107)^{m_w}$; $70.5 = (\sigma_o)_1 (0.0053)^{m_w}$;

$1.53 = 208.9^{m_w}$ and $m_w = 0.08$

Trial-and-error solution with Figure 7-9.

Assume $H_B = 135$; $d = 5.12$; $K_B = 0.465$; $S_u = 62.78$

For this annealed 1020: $S_u = 101(0.24/e)^{0.24} = 56.4$ ksi

for 20% CW $S_u = 56.4/0.8 = 70.5$

Since calculated S_u of 62.78 is less than 70.5; the value of H_B must be larger.

assume $H_B = 140$ $d = 5.05$ $K_B = .468$ $S_u = 65.52$
assume $H_B = 146$ $d = 4.95$ $K_B = .472$ $S_u = 68.9$
assume $H_B = 149$ $d = 4.90$ $K_B = .473$ $S_u = 70.5$

Therefore $H_B = 149$

Mechanical Properties 243

REFERENCES

1. Cottrell, A.H., *The Mechanical Properties of Matter,* John Wiley and Sons, New York 1964.
2. Datsko, J., *Material Properties and Manufacturing Processes,* John Wiley and Sons, New York 1967.
3. Everhart, J.L., *Mechanical Properties of Metals and Alloys,* National Bureau of Standards C447, Washington, D.C. 1943.
4. O'Neill, H., *Hardness Measurement of Metals and Alloys,* Chapman & Hall, London 1967.

STUDY PROBLEMS

7-1. A certain hot-rolled AISI 1020 steel has the following properties:
$S_y = 39$ ksi, $S_u = 66$ ksi, $\sigma_0 = 115$ ksi, $m = 0.22$, $A_r = 63\%$

   a. Plot the tensile yield strength and tensile ultimate strength vs. percent cold work on cartesian coordinates.
   b. Plot the tensile yield and tensile ultimate strengths vs. true strain (of cold work) on log-log coordinates.

7-2. A 0.505 in. dia. × 2 in. gage length standard threaded tensile specimen of annealed 1018 steel was loaded and strained until the minimum diameter was 0.340 in. During this loading, it was established that this material has a yield point of 35,000 psi and a strain-hardening equation of $\sigma = 105,000\,\varepsilon^{0.25}$. The specimen was unloaded before it fractured. (The published A_r value of annealed 1018 steel is 60% to 65%.)

   a. If this same piece were reloaded, what would be the new maximum load?
   b. If, instead of reloading immediately, the piece were machined to a uniform diameter of 0.300 in. over the entire length of the original

244   Chapter 7

gage length before retesting, what would be the new maximum load?

7-3. Nearly all common materials have a percent elongation less than their percent reduction of area. Is it possible for any material to have a percent elongation greater than its percent reduction of area? Explain.

7-4. A new material is developed, but because of the small amount available, a subsize tensile specimen having a 0.505 in. diameter and 1 in. gage length is used to determine its mechanical properties. The resulting elongation is 30% and area reduction is 30%. Estimate these two values if a 0.505 in. diameter and 2 in. gage length specimen are used.

7-5. A certain annealed nonferrous material has the following properties:
$S_u = 90,000 \text{ psi}, \quad S_y = 40,000 \text{ psi}, \quad A_r = 80\%$.

If this material is cold-worked 60% by axial tension, what will be the A_r of a tensile specimen machined from the cold-worked bar in the axial direction?

7-6. a. What amount of cold work is required to produce a natural strain of 6? Of 10?
b. What is the initial length of a bar required to have (after compression) a length of 1 in. with a natural strain of 10?

7-7. 1020 annealed steel has a microstructure that consists of 3/4 ferrite and 1/4 coarse lamellar pearlite. On this basis, it has a S_u of 60 ksi, m = 0.25, and H_B of 120 kg/mm². A 1/8" thick piece of this material is to be tested for a verification of its Brinell hardness. Since the piece is only 1/8" thick, a 500 kg load instead of the usual 3000 kg load is used. What should the Brinell hardness number be?

7-8. 1020 annealed steel has a 500 kg H_B of 100. What is its Rockwell B hardness?

7-9. A given heat of 1020 annealed steel has S_y = 30 ksi, S_u = 60 ksi, m = 0.25, A_r = 73%, H_B = 120 kg/mm². How hard will it be after 10% cold work. Note: you cannot use $(H_B)w = H_B/1 - wf$. Assume the cold-worked material has the relationship, $\sigma = (\sigma_o)_1 \varepsilon^{m1}$. Show all calculations.

7-10. The following mechanical properties are given for metal X:

Annealed

$S_y = 23\,\text{ksi}$, $S_u = 40\,\text{ksi}$, $E = 20 \times 10^3\,\text{ksi}$,

$El = 35\%$, $A_r = 30\%$, $H_B = 120\,\text{kg}/\text{mm}^2$

After cold work

%CW	5	10	20	30
S_y	64	79	97	109
S_u	84.2	89	100	110

Are all of the properties listed valid? If not, which are not valid. Show all calculations.

7-11. A recent ASTM research report lists the following properties for commercially pure titanium.

Temp.	S_u	S_y	A_r	2" El	σ_o	m
18°C	67	50	45	26	105	0.15
200°C	41	30	63	41	72	0.21
300°C	33	20	75	39	55	0.18

a. Are the 300°C properties listed consistent with each other?
b. Estimate the 500 kg Brinell hardness at 300°C. Show all calculations.
c. What alternating stress on a flat fatigue specimen will result in failure at 10^3 cycles at 300°C?

7-12. A steel forging in the suspension system of a foreign car failed and a product liability suit followed. The part was sent to a major industrial testing laboratory for examination and the following results were reported. A chemical analysis indicated 48.5 points of carbon and no significant amount of alloy, so it could be 1045 or 1050 steel. A metallurgical examination at 100× indicated it was 100% speroidite. Tensile tests gave the following results: $S_u = 106.6\,\text{ksi}$, $S_y = 97.4\,\text{ksi}$; El in 2" =

21 1/2%, A_r = 54%. Seventeen hardness readings ranged from 68 to 72 RB, with an average of 69.5. Are all of the data consistent?

7-13. A stainless steel flat spring is needed. The designed life expectancy is 1000 cycles. Compute the maximum alternating stress that can be tolerated if the material selected is:

a. 18-8 stainless steel annealed at 1800°F.
b. 17-4 PH stainless steel aged at 1100°F.

Use the data in Table B-2 of the appendix (p. 000). Compare it with the materials' S_u. Repeat the problem for 1,000,000 cycles.

7-14. Estimate the value of alternating stress that will cause a fatigue failure in 1000 cycles for the following metals:

a. 1020 HR steel having the following properties: S_y = 39 ksi; S_u = 66 ksi; σ_o = 115 ksi; m = 0.22; ε_f = 0.90
b. The same steel after cold rolling 20% (in the rolling direction).

Chapter 8

DEFORMATION STRENGTHENING AND FORMABILITY

INTRODUCTION

Cold formed parts are used extensively and increasingly in the manufacture of automobiles, airplanes, home appliances, and a wide variety of mechanical devices. Most of the common forming processes induce low-cycle strains of large magnitude in the fabricated part. For example, upsetting operations, which produce axial compressive strains, are performed on cold-drawn bars that have been subjected to prior axial tensile strains during drawing.

In the deep drawing of cylindrical cups, the metal on the outside surface near the top of the cup receives three cycles of strain during the drawing operation. First, as the metal is "drawn in" radially toward the die radius, it is subjected to a tensile strain in the radial direction. Then, as the metal is bent over the die radius, the inner fibers, which become the outer surface of the cup, are strained in compression. Finally, as the metal leaves the vertical tangency point of the die radius, it is unbent to form the straight wall of the cup and is subjected to tensile strains during this operation. If the cup is formed from cold-rolled strip, there is a fourth cycle of strain. And, if the cup wall is "ironed" during its last forming operation, there is an additional cycle of strain.

Since the mechanical properties of the material are altered by the forming operations, it is necessary for the design engineer to be able to calculate the strength of the material in the fabricated part. Otherwise, the mechanical reliability of the part may be poor, and the weight and the cost of the part may not be as low as they could be.

The traditional experimental method of determining the strength of a part by testing a prototype is too expensive to be widely used. Furthermore, it is not "engineering-wise" compared to the sophisticated design achievements being made. It is as desirable for the design engineer to be able to calculate the strength of a cold-formed part as it is to be able to calculate the stresses in it.

A material that has been cold worked has mechanical properties that are different in both the sense of the strength (tension or compression) and in the direction within the part (longitudinal or transverse). Therefore it is necessary, when referring to the strength of a cold-formed part, to specify both the sense and the direction of the strength that is being studied. In addition, when making a strength analysis as discussed in the following section, it is necessary to include the prior strain history at the location where the strength is being studied.

The traditional manner of presenting the yield strength of cold-worked materials without being this specific is completely inappropriate in today's sophisticated design age, where product reliability and liability are so important. Nearly all of the published data that show the variation in yield strength or tensile strength as a function of cold work still do not indicate either the sense or the direction of the strength. Some of this data are presented in the form of graphs or curves, with the percent cold work as the abscissa and the strength as the ordinate. More often, the data is presented in tabular form, as illustrated in Table 8-1, which is taken from one stainless steel manufacturer's data sheet.

Table 8-1. Effect of Cold Work on Mechanical Properties
(304 Stainless Steel)

% Cold reduction	Yield strength		Tensile strength		Elongation
	ksi	MPa	ksi	MPa	% in 2 in.
0	32.6	225	86.3	595	54.5
10	70.0	483	97.9	675	36.5
20	96.2	663	113.2	781	24.0
30	118.5	817	131.1	904	15.5
40	135.0	931	145.7	1005	12.0
50	145.0	999	158.3	1091	6.0

Deformation Strengthening and Formability   249

Nothing is stated about the mechanical properties in terms of the sense of the strength or its orientation in the part. How could an engineer use this information to design a shaft of 304 stainless steel that will have one of its ends upset by 30% cold work to form a gear blank? Since the load on a gear tooth is applied in the transverse or circumferential direction, the strength in the transverse direction is the appropriate one to use. In this particular case, is either the tensile yield strength or the compressive yield strength in the transverse direction equal to the 118.5 ksi listed in the table?

In order for cold-formed parts to be mechanically reliable, the design engineer must be able to calculate the actual strength at the critical locations in the part and make certain that the strength is greater than the stress by the appropriate factor of safety. Thus it is necessary to consider both the tensile and the compressive strengths. And the design engineer must know the values of the strength in whatever direction in the part the induced stresses act. It is not sufficient to base the design of a cold-formed part on the tensile values of the strength in only the longitudinal direction.

In addition, to ensure the reliability of a mechanical component, the design engineer must specify, and have control over, the important details of the actual fabrication processes that are used to make the part. For example, the engineer designing a hat section as shown in Figure 8-1 can no longer simply state on the blueprint: "cold form from x" thick sheets." If he/she designs the part on the basis of the material having a certain strength in the legs of the hat section, then he/she must also indicate whether the section should be formed on a press brake or a roll-forming machine, or in a draw die on a press. If it is formed in a draw die, as illustrated in Figure 8-2, then the designer must also specify the die radius inasmuch as it has a significant effect on the strength of the formed part. This is illustrated in the example at the end of this chapter. Likewise, an engineer designing an arc-welded joint must specify or otherwise have control over the welding variables of current, voltage, and welding velocity; otherwise, reliability cannot be achieved.

The mechanical design engineer of today has been well educated in the techniques of stress analysis. The tensile properties of a material are not required in the typical stress analysis problem since only stresses and strains are calculated. However, the calculated stresses are valid only if they are less than the material's yield strength. The design engineer can obtain information concerning yield strength of materials from handbooks, material manufactuerers' literature, or even from a materials engineer. Unfortunately, the data listed in the handbooks are for the material in its original condition as a bar or plate, whereas the relevant mechanical properties that are needed to make a valid stress analysis should be for a finished part.

This chapter introduces new concepts and relationships which make it possible for the design or manufacturing engineer to predict the strength of cold-

250   Chapter 8

worked metals. The calculations are based on the metal's original mechanical properties and the effect that mechanical processing (plastic deformation or cold work) has on those properties. These concepts, theories and relationships comprise a new engineering discipline that the author refers to as *strength analysis*, the importance of which is now being recognized in industry. Presented first is a new method for designating strength that includes both directionality and sense. Then the techniques for calculating the numerical values of the strength in all directions in fabricated parts are discussed.

DATSKO'S STRENGTH DESIGNATION

After several years' experience teaching this subject, the author has developed a very simple and useful notation for designating the strength of a fabricated part[1]. In this notation, strength is represented by the capital letter S followed by four subscript letters with parenthesis around the S and first letter. A typical example of this notation is: $(S_u)_{tLc}$, which means "the ultimate tensile strength in the longitudinal direction after prior axial compression."

To explain this designation, it is convenient to replace the letter subscripts with sequential numbers. In this form the standard designation is $(S_1)_{234}$. The first subscript indicates the *type* of strength and may be any of the following lower case letters: y, for yield; u, for ultimate; f, for fracture; or e, for endurance limit.

The second subscript refers to the *sense* of strength and it may be any of the following three lower case letters: t, for tensile; c, for compression; or s, for shear. In a following section where shear and torsional deformations are discussed, it is demonstrated that the concept of shear strength is not necessary since the tensile and compressive strengths are sufficient to define a material's strength.

The third subscript, in capital letters, indicates the *direction* or *orientation* of the test specimen within the formed part. L is used for a specimen whose axis is parallel to the direction of the principal deforming force or strain. For symmetrical cross sections such as rounds or squares, the designation is T, for transverse; R, for radial; or C, for circumferential. These latter three are all equivalent. In the case of plates or rectangular bars, two transverse directions must be specified: the long transverse can be indicated by B (breadth); the short transverse by H (thickness).

---

[1] Datsko, J. Material Properties and Manufacturing Process, John Wiley and Sons, New York 1966, p. 342

Deformation Strengthening and Formability 251

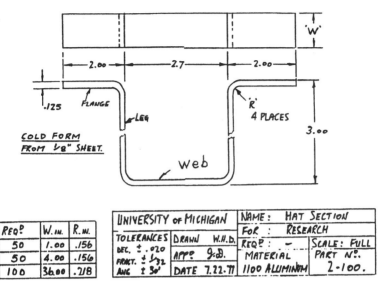

Figure 8-1. Blueprint of a hat section.

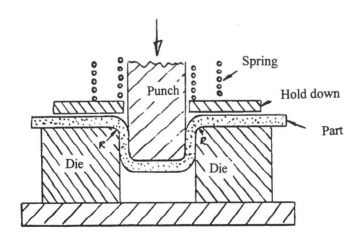

Figure 8-2. Draw die to form a hat section.

The fourth subscript refers to the *sense of the last prior strain in the axial direction of the test specimen,* and not in the direction of the deforming force. The notation here is the same as for the second subscript; that is: t, for tension; c, for compression; or s, for shear. However, if there has not been prior strain, the subscript o should be used. For a non-cold-worked or heat-treated isotropic material, the second, third and fourth subscripts can be deleted.

To generalize, the strength of a cold-worked material can be represented as $(S-)w$ and that of the original non-cold-worked material as $(S-)o$.

A second example is given to illustrate this designation. Consider the notation $(S_y)_{cTc}$, which means the compressive yield strength in the transverse direction after compressive strain in the transverse direction (equivalent to an extension in the longitudinal direction).

Four strength designations (illustrated in Figure 8-3) are needed to completely specify the strength in bars or plates that are upset (compressed in the longitudinal direction). The tensile and compressive strengths in the longitudinal direction are denoted as $(S-)_{tLc}$ and $(S-)_{cLc}$ respectively. The dash can be replaced by any of the symbols discussed above that can be used for the first subscript. The tensile and compressive strengths in the transverse direction are expressed as $(S-)_{tTt}$ and $(S-)_{cTt}$ respectively. In this case, the fourth subscript t indicates that the prior strain in the transverse direction was tensile.

Four strength designations are also needed for a bar that has been axially (longitudinally) stretched. In the longitudinal direction, the tensile and compressive strengths are $(S-)_{tLt}$ and $(S-)_{cLt}$, respectively. Similarly, in the transverse direction they are $(S-)_{tTc}$ and $(S-)_{cTc}$. Again, the last subscript c indicates that the prior strain in the transverse direction was compressive.

In the case of rolling or bending a plate (width $\geq 10 \times$ the thickness), six designations are necessary. There are both tensile and compressive strengths in each of the three perpendicular directions: longitudinal, L; long transverse, B; and short transverse, H. Thus the six designations are: $(S-)_{tLt}$ and $(S-)_{cLt}$; $(S-)_{tHc}$ and $(S-)_{cHc}$; $(S-)_{tBo}$ and $(S-)_{cBo}$. The last subscripts are t for the longitudinal direction since the length increases; c for the short transverse since the thickness is reduced; and o for the long transverse since the width remains constant and the strain in the long transverse direction is zero.

In the case of torsional or shear deformation, it must be recalled that there are induced tensile strains in one direction and compressive strains in another direction. It is simply necessary to determine the amount and direction of these two strains with respect to the axis of the bar. By this means, the tensile and compressive strengths in any direction can be calculated.

Deformation Strengthening and Formability 253

When the desired strength is at some angle other than 0° or 90°, that is, other than longitudinal or transverse, then a simple linear interpolation of the strength on the basis of the 0° and 90° values is required. Thus if a flat bar is uniaxially deformed in tension, and if the tensile yield strength at an angle of 30° to the longitudinal direction is desired, then the strength is calculated as follows:

$$(S_y)_{t\,30} = 2/3(S_y)_{tLt} + 1/3(S_y)_{tTc}$$

Likewise, the compressive yield strength is calculated as follows:

$$(S_y)_{c\,30} = 2/3(S_y)_{cLt} + 1/3(S_y)_{cTc}.$$

The techniques for calculating the numerical values for these strengths is presented in the following section.

DATSKO'S APPARENT RULES OF STRAIN STRENGTHENING

The uniaxial stress-strain characteristics of a material are determined by means of either a tensile or compressive test on a sample of the material. For most non-cold-worked materials, the results of the two tests are identical. In order to simplify the use of plastic stress-strain data it is beneficial to express the data mathematically. The two most commonly used expressions are:

$$\sigma = \sigma_0 \varepsilon^m \text{ and } \sigma = A + B\varepsilon$$

where σ is the true stress and ε is the true strain defined as the natural logarithm of the ratio of final to initial lengths. σ_0, m, A, and B are experimentally determined constants.

The author has found the first expression to be more reliable over a wide range of materials and it will be used exclusively in the following concepts and rules. However, these concepts and rules are general in that any valid mathematical expression of the material's stress-strain characteristics can be employed. It should be pointed out that the first expression is frequently written with other symbols, such as:

$$\sigma = K\varepsilon^m \text{ or } \sigma = K\varepsilon^n \text{ or } \sigma = A\delta^m$$

254   Chapter 8

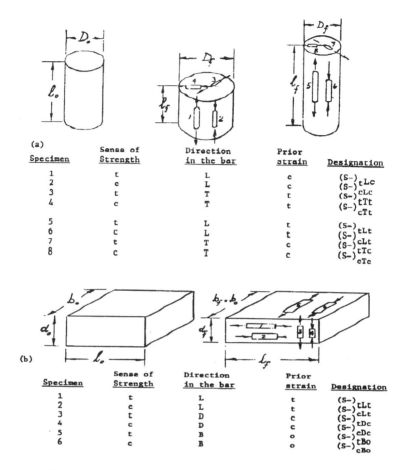

Figure 8-3.   Strength designation after (a) uniaxial deformation; (b) rolling a plate.

Since the uniaxial stress-strain curve of a material is also the locus of yield strengths of the material after prior tensile deformation,[2,3] its mathematical expression, $\sigma = \sigma_0 \varepsilon^m$, is really a strain-strengthening equation. Likewise it is more appropriate to call m the strain-strengthening exponent rather than the

---

[2]Datsko, J. Material Properties and Manufacturing Process, John Wiley and Sons, New York, 1966, p. 25

[3]Tabor, D. The Hardness fo Metals, Oxford University Press, London, 1951.

strain-hardening exponent; and σ_0 the strength coefficient. The numerical value of m varies between 0 and 0.7 for the entire range of metals.

The rules of strain strengthening can now be presented on the basis of the above notation of strength, and the stress-strain characteristics of materials. They are referred to by the author as "apparent rules" because they portray what the mechanical properties of a cold-worked material appear to be on the basis of a broad range of experimental studies. They have been developed by the author during the past 40 years as a means of presenting in a systematic manner the results of a large number of individual research projects concerning plastic deformation and mechanical properties, including cyclic axial deformation of cylinders; cyclic deformation of large cubes in the three axial directions; bending and unbending of flat specimens; cyclic torsional deformation of cylinders; shearing of blanks and strips; deep drawing of channel sections and cylindrical cups; wire drawing; forward and back extrusion; and cold rolling. A variety of tensile, compressive, fatigue, and hardness tests were conducted on the cold-worked metals in these studies.

Most of the author's early experimental work was done on single phase metals having a face-centered cubic lattice structure. More recently, tests have been conducted on several different steel compositions. In those tests where strain aging was avoided, the experimental data have been in agreement with the concepts presented below. These relationships show the effect of cold work alone on the strength of the material. They are valid only when no other strengthening mechanism such as precipitation or recrystallization occurs after the cold-working process.

Rule 1. Strain strengthening is a "bulk" mechanism. Even a deformation load that is applied in only one direction causes strengthening in all directions. For example, in the cold rolling of a plate, even though there is no increase in the width of the plate, both the tensile and compressive yield strengths in the width direction are increased by approximately 50-500%.

Rule 2. The maximum deformation that can be given a material during the forming of a part is that deformation which induces a tensile strain that is numerically equal to the fracture strain of that material when tested in the same direction and identical environment. In other words, fracture will occur when the induced tensile strain ε_t is equal to the fracture strain ε_f. This has been referred to by the author as *the failure theory for plastic working*[4]. However, from a practical approach, a few forming operations are terminated when the induced tensile strain is equal to the ultimate load strain, which is also

---

[4]Datsko, J. Material Properties and Manufacturing Process, John Wiley and Sons, New York, 1966, p. 315.

numerically equal to the strain strengthening exponent. This is true for those forming operations in which the condition known as necking occurs.

Some forming operations are a combination of stretching and bending, and in these cases the "failure" occurs at some value of strain between ε_u and ε_f that depends on the ratio of bending to stretching strain.

Rule 3. The numerical value of ε_w in any one deformation cycle that should be used in calculating the equivalent strain ε_{qu} is the largest of the three strains present at any location. It is usually the longitudinal or axial strain. For example, if a cube is compressed in the Z direction to a strain of –0.2, the strains in X and Y directions are each +0.1. The numerical value of strain that is used for ε_w in this case is 0.2, even when one is going to calculate the strength in the transverse direction.

The equivalent strain[5], ε_{qu}, is the value of the plastic (ulastic) strain after any combination of cold-work strains that can be used in the strain-strengthening relationship $\sigma = \sigma_o \varepsilon^m$ to calculate the strength of the deformed metal. The numerical value of ε_{qu} is calculated from the relationships given in Table 8.2.

Rule 4. In determining the value of ε_{qu}, the ε_{wi}'s are added in order of decreasing numerical value rather than in chronological order, as defined in Table 8.2, and without regard to their sign. Thus for a strain sequence of +0.10, –0.20, +0.15, –0.05, the value of ε_{qus} is 0.20 + 0.15/2 + 0.10/3 + 0.05/4 or 0.32, and the value of ε_{quo} is 0.20/2 + 0.15/3 + 0.10/4 + 0.05/5 or 0.19. The subscript s is added to ε_{qu} when the equivalent strain is to be used to calculate the strengths that have the 2nd and 4th subscripts in the strength designation of the same sense. That is, tensile yield strength after tensile strain or compressive yield after compressive strain. The subscript o is used when the strengths to be calculated have the opposite sense as that of the prior strain. This method of adding the strains gives the greatest effect to the largest strain, regardless of when in the actual strain sequence it occurs.

A cycle of strain occurs when there is a reversal in the sense of strain, or if there is a change from or to a zero strain, in any direction. For example, if a bar is cold drawn (extended in the axial direction) to a strain $\varepsilon_1 = +0.15$ and then a portion of it is compressed to a strain $\varepsilon_2 = -0.20$, the sequence of strains

---

[5]Borden, M. P. Multidirectional Tensile Properties of Materials Subjected to Large Cyclic Strains, Doctoral Thesis, University of Michigan, 1975.

is +0.15, −0.20, and there are two cycles of strain. However, if the cold-drawn bar has a portion of it forward extruded (extension in the axial direction again) to a strain of $\varepsilon_2 = +0.20$, then the strain sequence is +0.15, +0.20, or simply +0.35, and there is only one cycle of strain.

For only one cycle of strain, $\varepsilon_{qus} = \varepsilon_w$ and $\varepsilon_{quo} = \varepsilon_w / 2$.

Rule 5. The tensile strength of a ulasticly deformed metal is determined by means of the two equations:

$$(S_u)_w = (S_u)_o e^{\varepsilon_{qu}} \quad \text{for } \varepsilon_{qu} \leq m \tag{8-1}$$

$$(S_u)_w = \sigma_w = \sigma_o (\varepsilon_{qu})^m \quad \text{for } \varepsilon_{qu} \geq m. \tag{8-2}$$

Rule 6. The yield strength of a ulasticly deformed metal is determined by the equation

$$(S_y)_w = \sigma_o (\varepsilon_{qy})^m \tag{8-3}$$

where ε_{qy} is the equivalent strain that is used to calculate the yield strength, as defined in Table 8-2.

The yield strengths are divided into four groups. For any given strain sequence, the numerical values of all the yield strengths in any one group are equal. The largest yield strengths are in group 1, and the smallest yield strengths are in group 4.

In the elementary treatment of the relationship of yield strength to the amount of cold work based on the stress-strain equation $\sigma = \sigma\varepsilon^m$, it is stated that $(S_y)_w = \sigma_w = \sigma_o(\varepsilon_w)^m$. One of the restrictions implicit in this latter equation is that the yield strength $(S_y)_w$ be of the same sense and direction as the deforming strain, that is, longitudinal tensile yield after prior longitudinal tensile strain or else longitudinal compressive yield strength after longitudinal compression This equation is based on the approximately true assumption that the new yield load of an interrupted tensile test is equal to the prior maximum load on the specimen. On the basis of this assumption, the value of ε_{qys} would be equal to ε_{qu}, which in turn is equal to ε_w.

The equation for ε_{qys} in the group 1 strength designation was arrived at as a means to improve the accuracy of the above mentioned elementary approach. For most metals that are subjected to an interrupted tensile or compressive test, the new 0.2% offset yield load is a little lower than the prior

Chapter 8

Table 8-2. Equations for Strength Calculations of Ulasticly* Deformed Metals

1. $(S_u)_w = \sigma_w$ for $\varepsilon_{qu} > m$
2. $(S_u)_w = (S_u)_o e^{\varepsilon_{qu}}$ for $\varepsilon_{qu} < m$
3. $(S_y)_w = \sigma_o(\varepsilon_{qy})^m$

Group	Strength Designation	ε_{qu}	ε_{qy}
1.	$(S-)_{cLc}$		
	$(S-)_{tLt}$		
	$(S-)_{tBo}$	$\varepsilon_{qus} = \sum_{i=1}^{n} \frac{\varepsilon_{wi}}{i}$	$\varepsilon_{qys} = \frac{\varepsilon_{qus}}{1+0.2\varepsilon_{qus}}$
	$(S-)_{cBo}$		
	$(S-)_{cHc}$		
2.	$(S-)_{tTt}$		
	$(S-)_{cTc}$	$\varepsilon_{qus} = \sum_{i=1}^{n} \frac{\varepsilon_{wi}}{i}$	$\varepsilon_{qys} = \frac{\varepsilon_{qus}}{1+0.5\varepsilon_{qus}}$
3.	$(S-)_{tTc}$	$\varepsilon_{qus} = \sum_{i=1}^{n} \frac{\varepsilon_{wi}}{i}$	**
	$(S-)_{cTt}$		
4.	$(S-)_{cLt}$	$\varepsilon_{quo} = \sum_{i=1}^{n} \frac{\varepsilon_{wi}}{1+i}$	$\varepsilon_{qyo} = \frac{\varepsilon_{quo}}{1+2\varepsilon_{quo}}$
	$(S-)_{tLc}$		
	$(S-)_{tHc}$		

*Ulastic deformation is plastic deformation below a material's recrystallization temperature.

** $(S_y)_{tTc} = (S_y)_{cTt} = 0.9(S_y)_{tTt}$ or $0.9(S_y)_{cTc}$.

ε_{qus} = equivalent ulastic strain when the prestrain sense is the same as sense of strength.

ε_{quo} = equivalent ulastic strain when the prestrain sense is opposite to the sense of strength.

Deformation Strengthening and Formability

maximum load. This slightly improved equation gives a value of $\varepsilon_{qys} = \varepsilon_w / (1 + 0.2\,\varepsilon_w)$ and a more reliable relationship for $(S_y)_{tLt}$ or $(S_y)_{cLc} = \sigma_o \{\varepsilon_w / (1 + 0.2\,\varepsilon_w)\}^m$, which is a little lower than the true stress σ_w that was on the specimen just prior to the unloading.

EXAMPLE OF STRENGTH ANALYSIS

The application of the above concepts and relationships can best be illustrated by considering a typical strength analysis problem. Consider a 2.5" diameter head that is cold upset on the end of a 2" diameter shaft prior to machining splines on the enlarged end. To save some machining, a cold-drawn bar rather than an annealed one is selected. Assume that in the cold-drawing operation a 2 1/4" diameter bar is reduced to the 2" diameter. The material selected is a 304 stainless steel having a yield strength of 35 ksi, a tensile strength of 86 ksi, $\sigma_o = 200$ ksi, m = 0.50, and $\varepsilon_f = 1.7$.

Since the force acting on the splines is in the circumferential, or transverse, direction, it is necessary to know both the tensile and compressive strengths. Only the yield strengths and tensile strength will be calculated in this example.

The cold drawing strain is calculated as: $\varepsilon_1 = 2\ln 2.25 / 2.00 = 0.24$ since the definition of strain is $\varepsilon = \ln l_f / l_o = \ln A_o / A_f = 2\ln D_o / D_f$.

The upsetting strain is $\varepsilon_2 = 2\ln 2.00 / 2.50 = -0.45$.

According to the notations and equations in Table 8-2 the equivalent ulastic strains are calculated as:

$$\varepsilon_{qus} = 0.45 + 0.24 / 2 = 0.57$$
$$\varepsilon_{quo} = 0.45 / 2 + 0.24 / 3 = 0.30$$

The transverse tensile yield strength is calculated as:

$$(S_y)_{tTt} = \sigma_o (\varepsilon_{qys})^m$$

where $\varepsilon_{qys} = 0.57 / (1 + 0.5 \times 0.57) = 0.44$

$$(S_y)_{tTt} = 200(0.44)^{0.5} = 133 \text{ ksi}$$

The transverse compressive yield strength is calculated as:

$$(S_y)_{cTt} = 0.9(S_y)_{tTt} = 0.9 \times 133 = 120 \text{ ksi}$$

However, if the properties are measured experimentally in the longitudinal direction, they will more closely resemble the following calculated values.

$$(S_y)_{cLc} = 200\left(0.57 / (1 + 0.2 \times 0.57)\right)^{0.5} = 143 \text{ ksi}$$

$$(S_y)_{tLc} = 200\left(0.30 / (1 + 2 \times 0.30)\right)^{0.5} = 87 \text{ ksi}$$

which is nearly half the compressive value.

The tensile strength in the transverse direction is:

$$(S_u)_{tTt} = 200(0.57)^{0.5} = 151 \text{ ksi}$$

In the longitudinal direction the tensile strength is:

$$(S_u)_{tLc} = 86 \, e^{0.30} = 116 \text{ ksi}$$

which is considerably lower than the value for the compressive longitudinal yield strength.

DETERMINATION OF DEFORMATION STRAINS

In order to calculate the strength of the material after deformation, and also to predict when failure will occur during deformation, it is necessary to know the magnitude of the strain at all critical locations of the part. In some complex forming operations, it is impossible to calculate the magnitude of the strains on the basis of the geometry of the tools and the finished part. In these cases, it is necessary to devise a procedure by which the strains can be experimentally determined. Several techniques can be used. One of the most common methods is to etch a grid on the surfaces of the blanks before forming. Then the changes in the dimensions of the grid are measured during several stages of the forming operation and the maximum strains are thus determined. Two of the other methods are to measure the thickness and width of the part, or else to take hardness readings at various locations which are then compared to standard calibration curves of hardness versus strain.

Deformation Strengthening and Formability

There are many forming operations for which the magnitude of the strains at all locations can be calculated analytically with sufficient reliability on the basis of the geometry of the part and the tools. These are the only ones discussed in this chapter.

Axial Deformation

For simple stretching and upsetting with negligible friction, the axial strain is the natural logarithm of the ratio of final to original lengths. If the material is not in the cold-worked condition, does not have a significant amount of non-metallic inclusion of fibers, and has a cubic lattice structure, then the two perpendicular transverse strains are equal to one-half of the axial strain. In many sheet metals, the presence of inclusions, or a preferred orientation of the grain structure, will cause the two transverse strains to be unequal. However, in all forming operations the sum of the three principal strains at any location must equal zero since the volume does not change. Because of this constancy of volume, the axial strain is also equal to the natural logarithm of the ratio of cross-sectional areas.

Rolling Deformation

In the rolling of plates (width equal to or greater than 10 times the thickness), the width strain is zero. Thus the thickness strain and length strain are equal in magnitude but of opposite sense. The principal deforming strain during rolling appears to be tensile in the longitudinal (rolling) direction.

Bending Deformation

The key to the strain analysis during bending is the location of the instantaneous neutral axis; that is, the radial location of the arc length that is equal to the original length in the unbent part. It is also important to know the strain history of the metal that lies slightly below the mid thickness.

The popular concept of the location of the neutral axis during bending is that it moves inward to the radial position equal to the square root of the product of the inner fiber radius and the outer fiber radius. That is, $R_N = \left(R(R+h)\right)^{0.5}$, where R is the inner fiber radius and h is the thickness. The neutral axis is assumed to be at this location because it is equated to the neutral surface which is described in detail elsewhere (see Hoffman and

262   Chapter 8

Sachs[6], Mellor and Johnson[7] or R. Hill[8]). But the neutral surface is the radial position where the instantaneous stress is zero. That is, the material above the neutral surface is stressed in tension, while the material beneath it is stressed in compression. However, the instantaneous strain is not zero at this location. It is demonstrated below that the neutral axis (the location where the strain is zero) always remains at the mid-thickness of the plate.

One of the attributes of this concept of the neutral axis is that the positive outer fiber strains are numerically equal to the negative inner fiber strains.

Although this concept appears valid for stress analysis during bending, the author has found a serious fallacy in this model; namely, that the volume of the bent metal is significantly greater than the original volume. This conclusion can be verified analytically by simply comparing the original area A_o (ABCD) to the bent area $A_b(A'B'C'D')$ for a unit width plate, as shown in Figure 8.4. For example, if $R = h = 1$ and the radius of the neutral axis R_N is at $(R(R+h))^{0.5}$ (that is, the arc length along R_N is equal to the original length l_o), then the bent area is 6% greater than the original area. Since the volume of metal is constant during plastic deformation, the above model must be incorrect.

The location of the neutral axis based on the constancy of volume can be found as follows (and with reference to Figure 8.4).

By definition the arc length of the neutral axis l_N is equal to the original length l_o. But l_N is also equal to θR_N in the bent plate.

The original area: $A_o = l_o h$

By substitution: $A_o = \theta R_N h$

The bent area: $A_b = \pi\left((R+h)^2 - R^2\right)\theta/2\pi$

By equating A_o to A_b, the following relationship is obtained:

$$(R+h)^2 - R^2 = 2 R_N h \text{ and finally } R_N = R + h/2$$

---

[6]Hoffman, O. and G. Sachs, Theory of Plasticity for Engineers, McGraw-Hill, New York, 1953.
[7]Mellor, P.B. and W. Johnson, Plasticity for Mechanical Engineers, Van Nostrand, New York, 1962.
[8]Hill, R. The Mathematical Theory of Plasticity, Oxford University Press, New York, 1950.

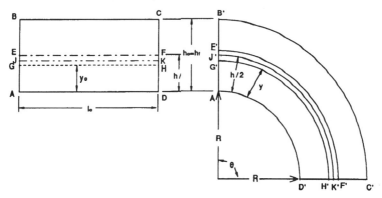

Figure 8-4. Model for the bending of a plate (w/h ratio greater than 10 where w is the width of the plate) when the edge effects can be neglected. (a) The original plate. (b) The plate after pure bending to inside radius R. $h_f = h_o$ and $w_f = w_o$. y is the location in the bent plate of the element y_o in the original plate.

Thus, it is apparent that during plastic bending of plates the instantaneous neutral axis remains at the mid-thickness of the plate. Since the value of the inner fiber strain is $\varepsilon_{if} = \ln(R/R_N)$ and the outer fiber strain is $\varepsilon_{of} = \ln((R+h)/R_N)$, the numerical value of the inner fiber strain is greater than the outer fiber strain.

This analysis of the location of the neutral axis has been verified experimentally by the author by two methods: (1) measuring the inner and outer fiber strains after bending, and (2) by conducting microhardness studies at both surfaces of bent metals.

Analysis of the strain history. Since the neutral axis remains at the mid-thickness during pure bending, all points below the mid-thickness are in compression and all points above the neutral axis are in tension. However, some of the metal that originally was below the mid-thickness (area JEFK) is displaced during bending to a position above the mid-thickness (area J'E'F'K'). These elements are first compressed and then stretched so that they have a final residual tensile strain. And there is one element, JK, below the mid-thickness that is first compressed and then displaced upward until it lies at the mid-thickness at which time its strain is zero.

Chapter 8

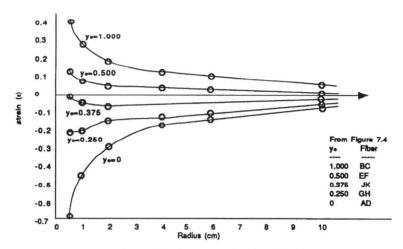

Figure 8-5. The strain history during bending for a 1 cm thick plate bent to a final radius of 0.5 cm.

The following mathematical relationships between the thickness of the plate (h), inside radius of the bend (R), and circumferential strains at any location can be derived with reference to Figure 8.5. Since the width strain for plates or sheets having a width-to-thickness ratio equal to or greater than 10 is zero, the thickness strain is numerically equal to but of opposite sense to the circumferential strain. These relationships are valid for all metals; only the resulting stresses and strength vary for the different metals.

1. The distance y_o for given values of R, h, and y is determined by equating the area AGHD to A'G'H'D'.

$$y_o = \frac{y(2R+y)}{2R+h} \qquad (8\text{-}4)$$

2. The distance y for given values of R, h, and y_o is obtained by rewriting the previous equation for y.

$$y = \left(R^2 + 2R \times y_o + hy_o\right)^{0.5} - R \qquad (8\text{-}5)$$

3. The strain at y for given values of R, h, and y is found as follows:

Deformation Strengthening and Formability 265

$$l_y = \theta(R+y)$$

and

$$l_o = l_N = \theta(R+h/2)$$

so

$$\varepsilon_y = \ln(l_y/l_o) = \ln\left(\frac{R+y}{R+h/2}\right) \tag{8-6}$$

4. The strain at y for given values of R, h, and y_o is obtained by substituting for y Eq. (8-6) the value given in Eq. (8-5). Thus:

$$\varepsilon_y = \ln\left(\frac{\left(R^2+2R\times y_o+hy_o\right)^{0.5}}{R+h/2}\right) \tag{8-7}$$

5. The radius that results in the maximum compressive strain for location y_o and thickness h is $(Ry_o)_m$. This relationship pertains to those values of y_o that first experience a compressive strain at the beginning of the bending operation but that subsequently experience a tensile strain. $(Ry_o)_m$ is found by differentiating Eq. (8-7) with respect to R and equating the result to zero. This gives:

$$(Ry_o)_m = \frac{hy_o}{h-2y_o} \tag{8-8}$$

6. The location of y_o for $\varepsilon_y = 0$ when R and h are given is the location before bending that moves up to the mid-thickness location after bending. When $\varepsilon_y = 0$ in Eq. (8-7),

$$\frac{\left(R^2+2R\times y_o+hy_o\right)^{0.5}}{R+h/2} = 1$$

and

$$y_o = \frac{\left(Rh+h^2/4\right)}{2R+h} \tag{8-9}$$

266   Chapter 8

7. The radius R when $\varepsilon_y = 0$ for given values of h and y_o is found by rewriting Eq. (8-9).

$$R = \frac{hy_o - h^2/4}{h - 2y_o} \tag{8-10}$$

8. The value of y_o that has zero slope for radius R and thickness h is obtained by rewriting Eq. (8-5). This value is designated $(y_o)_o$.

$$(y_o)_o = \frac{Rh}{2R + h} \tag{8-11}$$

9. The maximum compressive strain at y for given values of h and y_o occurs when the radius R is equal to $(Ry_o)_m$. This value of strain is designated as $(\varepsilon_y)_m$ and is found by substituting $(Ry_o)_m$ for R in Eq. (8-7).

$$(\varepsilon_y)_m = \ln \frac{\left((Ry_o)_m^2 + 2(Ry_o)_m y_o + hy_o\right)^{0.5}}{(Ry_o)_m + h/2} \tag{8-12}$$

Figure 8-5 is a graphical presentation of the strain history that occurs when a 1 cm thick plate is bent to a final inside radius of 0.5 cm. For values of y_o between 0.5 cm (mid-thickness) and 1.0 cm (outer fiber), the strains continuously increase in tension to their final values of 0.11 and 0.41, respectively. For values of y_o between 0.25 cm (location where the slope of the strain curve is zero) and $y_o = 0$ cm (inner fiber), the strains continuously increase in compression to their final values of –0.14 and –0.69, respectively. The ratio of the inner fiber to outer fiber strains is 1.68. This ratio decreases inversely with the bend radius. For example, when the radius is 2 cm (R/h = 2), the inner fiber strain is –0.22 and the outer fiber strain is 0.18, which results in a ratio of 1.22.

At the location where $y_o = 0.375$ cm (the line that finally becomes the mid-thickness when the radius is 0.5 cm), the strain first increases in compressive value and then is subjected to tensile strains such that the final strain is zero when the radius is 0.5 cm. At this location, the maximum compressive strain was –0.032 (equivalent to 3% plastic deformation) when the radius was 1.5 cm.

All elements between $y_o = 0.375$ and 0.5 cm first experience a compressive strain but end up with a final tensile strain. And all elements between $y_o = 0.25$ and 0.375 cm first undergo a compressive strain that is followed by a tensile strain of smaller value so that the final strain is still compressive.

Torsion and Shear Deformation

Many forming processes include torsional or shear deformations. One automobile manufacturer subjects torsion bars to large permanent torsional deformation as a means of increasing their shear strength. Shear or power spinning is a forming process in which a disc is deformed into a cone by means of shear deformation In the extrusion and bar-drawing processes, the material near the outside surfaces of the finished part has been subjected to both axial extension and shear deformation, while the material at the center of the bar has had only an axial extension.

To predict when failure will occur due to shear deformation during forming, or to predict the strength of the finished part, it is necessary to be able to determine the magnitude of the induced strains.

Figure 8-6 introduces a new method developed by the author to calculate the strains induced by large torsional deformations. The strains calculated by this method can be used to predict both the amount of deformation necessary to cause fracture and the resulting strengths after any amount of torsional deformation. This model does not make use of the concept of shear strain, $\gamma = r\theta / l$, that is used in the mechanics of deformable bodies where the shear strain is related to the tensile strain.

Instead, this concept of torsional deformation is based on the only definition of strain that the author has found to be needed to solve all of the problems in the area of strength analysis, namely, $\varepsilon = \ln\left(l_f / l_o\right)$. To apply this concept of strain to torsional deformation, consider the solid cylinder illustrated at the top of Figure 8-6. When the right end of the cylinder is twisted clockwise through some angle θ, point B moves to C. Point B is arbitrarily selected so that PB lies at some angle α above PA, which is an axial line. Angle α is selected such that the original line PB receives the maximum tensile strain at the completion of the torsional deformation. Of course, when α is zero, B and A coincide.

As shown in Figure 8-6, line PB is designated as l_o to indicate it is an original length. After twisting the right end through an angle θ, B moves to C

268 Chapter 8

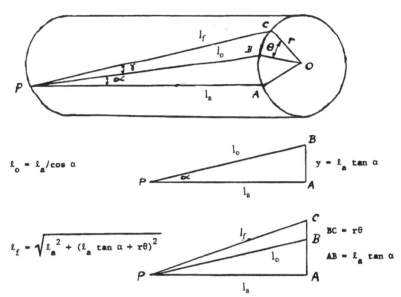

Figure 8-6. Determination of strains for large plastic torsional deformations.

and the length PC is identified as l_f to indicate that it is the final length. γ is the angle between l_f and l_o. In the top of the figure, l_o and l_f are helical lines.

In the first triangle beneath the cylinder, the surface PAB is represented as a planar surface. The length l_o can be expressed in terms of the angle and the axial length l_a. Thus $l_o = l_a / \cos \alpha$. Also, the distance AB, designated as Y, is equal to $l_a \tan \alpha$.

As shown in the bottom pair of triangles, when B moves C, the distance BB' is equal to $r\theta$. The final length l_f is the hypotenuse of the triangle PAC. Therefore $l_f = \left(l_a^2 + (l_a \tan\alpha + r\theta)^2 \right)^{0.5}$.

Since the definition of strain is $\varepsilon = \ln(l_f / l_o)$ the strain induced in PB as it moves to PC is:

$$\varepsilon = \ln \frac{\left(l_a^2 + (l_a \tan\alpha + r\theta)^2 \right)^{0.5}}{l_a / \cos\alpha} \qquad (8\text{-}13)$$

This can be simplified to:

$$\varepsilon = \ln \cos\alpha \left(1 + \left(\tan\alpha + r\theta/l_a\right)^2\right)^{0.5} \quad (8\text{-}14)$$

Thus, it is apparent that the strain induced in PB as the right end of the cylinder is twisted through an angle θ is a function of the angle α. In other words, the strain induced in the axial line PA is not the same as the strain induced in PB where α is any angle other than 0. It is possible to determine the value of the angle α that gives the largest strain ε for any amount of twist θ by differentiating the above equation for strain with respect to the angle α and equating the result to zero. This produces the equality $\tan 2\alpha = 2l_a/r\theta$. Solving this for α gives:

$$\alpha = 1/2 \tan^{-1}(2l_a/r\theta) \quad (8\text{-}15)$$

Thus when any cylinder of length l_a and radius r is twisted through some angle θ, it is the line PB that lies at the angle α above the axial line that has the maximum tensile strain induced in it. In this analysis, α is in degrees and θ is in radians.

Consider how this model predicts the strains for small deformations encountered in stress analysis problems dealing with elastic deformations. As θ approaches 0, the arc tan of infinity is 90° and therefore α is 45°. This agrees with the mechanics concept that the maximum tensile stresses and strains on the surface of a shaft subjected to torsion occur at an angle of 45° to the axis of the shaft.

But consider an extremely ductile material that can withstand enormous deformation. As θ approaches infinity, α approaches 0. Thus for extremely large deformations, it is an axial line, and not the 45° line, that undergoes the greatest tensile strain. And, of course, fracture will occur when this induced tensile strain is numerically equal to the fracture strain determined from a tensile specimen of this particular material.

This theory was verified at the same time that the author conceived it. The testing procedure was extremely simple. Since a supply of 1100-0 aluminum standard size (1/2" diameter × 2" gage length) threaded tensile specimens were available on the shelf in the laboratory it was decided to use them for the torsional deformation also. From previous tensile tests on this supply of aluminum specimens it was known that they had a fracture strain of approximately 2.3. The above mentioned theory implies that a 1/2" diameter × 2" long cylinder of this aluminum would require about 12 full revolutions of

Chapter 8

twist before it would fracture. Several specimens were subjected to torsional deformation while supported between centers on a lathe in the machine shop, with care to avoid axial constraints. The specimens did in fact fail by shear in eleven to twelve revolutions. Subsequently, tests were performed on a variety of ferrous and non-ferrous metals and all of them verified this theory of torsional deformation.

The following derived relationships are very useful in calculating the value of tensile strains induced during torsional deformation.

1. When r, l_a and θ are known, then

$$\alpha = 1/2 \tan^{-1}(2 l_a / r\theta) \quad \text{and} \quad (8\text{-}16)$$

$$\varepsilon = \ln(1 / \tan \alpha). \quad (8\text{-}17)$$

2. When r, l_a and ε are given, then

$$\alpha = \tan^{-1} e^{-\varepsilon} \quad \text{and} \quad (8\text{-}18)$$

$$\theta = l_a \left(e^{\varepsilon} - e^{-\varepsilon} \right) / r. \quad (8\text{-}19)$$

Figure 8-7 presents the method developed by the author to calculate the strains induced by large "pure" shear deformations. For pure shear, the thickness of the shear element (perpendicular to the plane of the element) does not change and the normal distance between the skewed sides (BC and AD in Figure 8-7) does not change. At the top of Figure 8-7 is the traditional model of shear deformation used in stress analysis problems. It is not satisfactory for large deformations.

The second model of shear deformation, for large deformations, has been developed by the author. The strains calculated by this model enable one to predict when failure will occur during shear deformation and also the strength of the material after any amount of shear deformation. In this model the element ABCD is deformed in shear by some amount that is defined by the angle ϕ so that it takes the shape AB'C'D. To determine the maximum tensile strain, a point E is placed on the line BC so that AE is at an angle α to the line AB, as shown in (a). The line AE is designated as l_o, the original length. After the deformation shown in (b) and defined by the angle ϕ, the point E is located at E'. The distance B'E' remains equal to BE. The line AE', designated l_f, is the final length of the original line l_o.

Deformation Strengthening and Formability 271

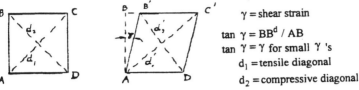

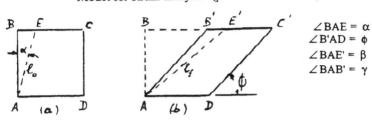

Figure 8-7. Maximum strain induced during shear deformation.

The definition of strain is: $\varepsilon = \ln(l_f/l_o)$. Since the breadth (thickness) does not change, $\varepsilon_b = 0$. Therefore, $\varepsilon_c = -\varepsilon_t$. That is, the compressive strain at any point is equal but of opposite sense to the tensile strain.

For pure shear the induced tensile strain is then

$$(\varepsilon_t)_i = \ln(l_f/l_o) = \ln(AE'/AE) \tag{8-20}$$

As shown in Figure 8-7, the angle α is a function of the amount of deformation given the element. That is, α depends upon the magnitude of ϕ. The relationship is derived as follows:

$$AE = AB/\cos\alpha = 1/\cos\alpha \quad \text{for} \quad AB = 1$$
$$BE' = BB' + BE = AB\cot\phi + AB\tan\alpha = \cot\phi + \tan\alpha$$
$$AE'/AE = \cos\alpha\left((\cot\phi + \tan\alpha)^2 + 1\right)^{0.5}$$
$$d(AE'/AE)/d\alpha = 0 \quad \text{gives}$$
$$\tan\alpha = \left(\left(\cot^2\phi + 4\right)^{0.5} - \cot\phi\right)/2$$

272   Chapter 8

so
$$\alpha = \tan^{-1}\left(\left(\cot^2\phi + 4\right)^{0.5} - \cot\phi\right)/2 \qquad (8\text{-}21)$$

or
$$\alpha = 0.5\tan^{-1}(2\tan\phi) \qquad (8\text{-}22)$$

or in terms of γ, (angle BAB')

$$\alpha = \tan^{-1}\left(\left(\tan^2\gamma + 4\right)^{0.5} - \tan\gamma\right)/2 \qquad (8\text{-}23)$$

The induced tensile strains can be calculated from the relationships

$$(\varepsilon_t)_i = \ln\cos\alpha\left(\left(\cot\phi + \tan\alpha\right)^2 + 1\right)^{0.5} \qquad (8\text{-}24)$$

or
$$(\varepsilon_t)_i = \ln\cos\alpha\left(\left(\tan\gamma + \tan\alpha\right)^2 + 1\right)^{0.5} \qquad (8\text{-}25)$$

In the cases were the induced strain ε or the angle α is given, the following relationships can be used.

$$\alpha = \tan^{-1}e^{-\varepsilon} \quad \text{also} \quad \alpha = 1/2\tan^{-1}(2\tan\phi) \qquad (8\text{-}26)$$

and
$$\varepsilon = \ln(1/\tan\alpha) \qquad (8\text{-}27)$$

and
$$\phi = \tan^{-1}\left(\tan\alpha/\left(1 - \tan^2\alpha\right)\right) \qquad (8\text{-}28)$$

FORCES, WORK, AND ENERGY DURING DEFORMATION

Often it is necessary to be able to calculate the forces needed to perform a forming operation. There is also considerable interest in energy-absorbing devices, particularly in the automotive field. Some of these devices utilize the plastic deformation of a piece of metal to absorb the energy. The following examples illustrate how the plastic stress-strain characteristics of a metal can be used to predict the deforming forces and the deformation energy.

Deformation Strengthening and Formability 273

As shown in Figure 8-8, the force to deform a part in uniaxial tension or compression varies with the amount of deformation. The instantaneous force is always a product of the instantaneous stress × the instantaneous area. That is, $F_i = \sigma_i A_i$, where $\sigma_i = \sigma_0 (\varepsilon_w)^m$. During compressive deformation, the force increases until fracture occurs. However, during tensile deformation, the force increases to a maximum value when the strain ε_w is equal to the ultimate load strain ε_u. Then necking begins and the force decreases until fracture occurs.

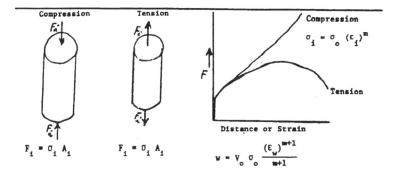

Figure 8-8. Deforming forces in tension and compression.

The work that is done to achieve this deformation is equal to the energy absorbed by the metal if the frictional work is negligible. The work per unit volume of the material to do the deformation is equal to the area under the $\sigma - \varepsilon$ curve for the material. That is:

$$W/V_0 = \int \sigma \, d\varepsilon \quad \text{and} \quad \sigma = \sigma_0 (\varepsilon_w)^m$$

$$W/V_0 = \int_0^{\varepsilon_w} \sigma_0 \varepsilon_w^m \, d\varepsilon$$

After integration the result is:

$$W = \frac{V_0 \sigma_0 (\varepsilon_w)^{m+1}}{m+1} \tag{8-29}$$

274   Chapter 8

This relationship is valid only when the entire volume of the metal is deformed uniformly. This is true for uniaxial compression: but it is true for uniaxial tension only for values of ε_w up to ε_u.

In the bar drawing and the extrusion processes shown in Figure 8-9, the forces are constant during the process except at the very beginning. The forces can be calculated for these operations by equating the work absorbed by the material, which is $W = A_o l_o \sigma_o (\varepsilon_w)^{m+1} / m+1$ to the work done on the material, which is $W = F \times l_o$ for extrusion and $W = F \times l_f$ for drawing.

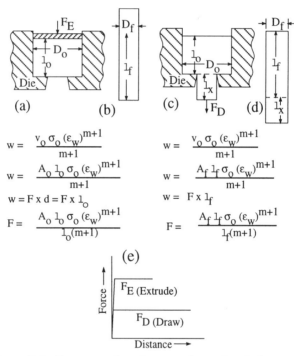

Figure 8-9. Forces, work and energy absorption during plastic deformation.

Sketches a and b in Figure 8-9 illustrate the before and after shapes during the extrusion process. Sketch c shows a billet of length l_o in a draw die with a gripping end l_x extending outside the die. Sketch d illustrates the drawn bar where length l_o is increased to l_f.

Deformation Strengthening and Formability   275

The force to extrude, neglecting friction and redundant work, can be estimated from the equation:

$$F_E = \frac{A_o \sigma_o (\varepsilon_w)^{m+1}}{m+1} \qquad (8\text{-}30)$$

Likewise the force to draw is:

$$F_D = \frac{A_f \sigma_o (\varepsilon_w)^{m+1}}{m+1} \qquad (8\text{-}31)$$

EXAMPLE PROBLEMS

8-1. A 1" long by 0.505" diameter cylinder is loaded in tension to a stress equal to $1.10 \times (S_u)_o$. It is then compressed to a 1" length. The material is soft nickel 200.

$(S_y)_o = 16.2$ ksi, $(S_u)_o = 72.1$ ksi, $\sigma_o = 150$ ksi,
$\varepsilon_u = 0.375$, $\varepsilon_f = 1.805$

Plot the load (k#) vs. l (inches) curve during these deformation cycles.

Solution:
$S_y = 16.2$; $S_u = 72.1$; $\sigma_o = 150$; $\varepsilon_u = 0.375$; $\varepsilon_f = 1.805$;
$l_o = 1.00$; $A_o = 0.200$; $(L_y)_o = 3.24$ k#

$\sigma_z = 1.10(S_u)_o = 1.1 \times 72.1 = 79.31 = 150(\varepsilon_z)^{0.375}$ $\varepsilon_z = 0.183$

Since $\varepsilon_z < \varepsilon_u$, it is not loaded up to the ultimate load which is
$L_u = 72.1 \times 0.2 = 14.42$ k#   and $l_u = 1.455$"
$A_z = A_o / e^{\varepsilon_z} = .2 / e^{0.183} = 0.167 \text{ in}^2$
$L_z = 79.31 \times 0.167 = 13.24$ k#   and $l_z = 1.200$"

Chapter 8

In compression, at the beginning $\varepsilon_{quo} = 0.183/2$

$$(S_y)_{cLt} = \sigma_o \left(\frac{\varepsilon_{quo}}{(1 + 2\varepsilon_{quo})} \right)^m = 150(.091/1.183)^{0.375} = 57.33$$

$$(L_y)_c = 57.33 \times 0.167 = 9.57 \, k\#$$

at end: $\varepsilon_{qus} = 0.183 + 0.183/2 = 0.275$

$$\sigma_z = \sigma_w = \sigma_o (\varepsilon_{qu})^m = 150(0.275)^{0.375} = 92.37 \, ksi$$

$$L_z = 92.37 \times 0.2 = 18.48 \, k\#$$

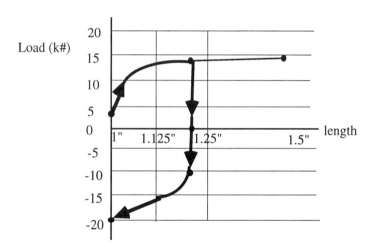

8-2. A heat of 1010 steel is in the form of 10 gage (0.1345") pickled and oiled HR sheets. It has the following tensile properties:
$S_y = 30 \, ksi$; $S_u = 50 \, ksi$; $m = 0.27$; $\varepsilon_f = 1.1$

  a. What load will a 1" wide strip of steel carry before yielding? What is the weight of a 1" wide by 10" long piece?
  b. If the 10 gage sheets are cold rolled to 12 gage (0.1046), what width of strip is needed to carry the same load without yielding? What is the weight of a 10" long piece at this new width and thickness?
  c. What is the ductility of the cold-rolled sheets in the longitudinal direction?

Deformation Strengthening and Formability 277

Solution:

a. $S_y = 30$ ksi; $A = 0.1345$ in.^{2}; $L_y = 30$ ksi $\times 0.1345$ in.$^2 = 4.035$ k#
    wt. $= r \times V = 0.283$ #/in.$^3 \times 0.1345$ in.$^2 \times 10$ in. $= 0.38$#

b. $\sigma_o = 50(e/0.27)^{0.27} = 93.3$ ksi $\varepsilon_w = \ln 0.1345/0.1046 = 0.252$
    $(S_y)tLt = 93.3(0.252/1+0.0504)^{0.27} = 63.5$ ksi
    $b = L_y / S_y \times h = 4.035/63.5 \times 0.1046 = 0.61$
    wt.$= 0.283 \times 0.1046 \times 0.61 \times 10 = 0.18$#
    % lighter $= (0.38 - 0.18)/0.38 = 53\%$

c. $(\varepsilon_f)_w = \varepsilon_f - \varepsilon_w = 1.1 - 0.252 = 0.848$
    $\varepsilon^{0.848} = 2.34 = 100/(100 - A_r)_w$
    $(A_r)_w = 57.3\%$ Note: $(A_r)_o = 66.5$

8-3. A 6" high by 3" diameter bar of annealed metal having $\sigma = 100$ ksi, m = 0.50, $S_y = 20$ ksi, $\varepsilon_f = 1.1$ is compressed to a height of 4.8"

   a. Neglecting friction, what is the final compressing load?
   b. What is the longitudinal tensile ultimate strength after compression?
   c. What is the transverse compressive yield strength?

Solution:

a. $\varepsilon_w = \ln(6/4.8) = 0.223$
    $\sigma_i = 100(0.223)^{0.5} = 47.22$ ksi
    $L_i = \sigma_i A_i$
    $A_i = A_o \times e^{0.223} = 8.84$ in.^{2}
    $L_i = 47.22 \times 8.84 = 417$ k#

b. $(S_u)_{tLc} = (S_u)_o e^{\varepsilon quo} = 100(0.5/e)^{0.5} e^{0.111} = 47.92$ ksi

c. $(S_y)_{cTt} = 0.9(S_y)_{tTt} = 0.9\sigma_o(\varepsilon_{qus}/(1+-0.5\varepsilon_{qus}))^m$

Chapter 8

$$= 0.9 \times 100(0.223/(1+0.111))^{0.5} = 40.3 \text{ ksi}$$

8-4. A heat of 70%Cu - 30%Zn brass annealed at 1200°F for 1 hour has the following tensile properties: $S_y = 11$ ksi; $S_u = 44$ ksi; $\sigma_o = 110$ ksi; $\varepsilon_u = 0.62$; $\varepsilon_f = 1.1$.

A 2" square × 3" long specimen was given the following deformation in the z (3" long) direction: (1) $\varepsilon_w = -0.11$; (2) $\varepsilon_w = +0.11$; (3) $\varepsilon_w = -0.12$; (4) $\varepsilon_w = +0.14$.

    a. Calculate the compressive yield strengths in the x, y, and z directions.
    b. Calculate the tensile ultimate strengths in the x, y, and z directions.
    c. Calculate the tensile yield strength in the x-z plane at an angle of 60° from the z direction.

Solution:

$\varepsilon_{qus} = 0.14 + 0.12/2 + 0.11/3 + 0.11/4 = 0.264$

$\varepsilon_{quo} = 0.14/2 + 0.12/3 + 0.11/4 + 0.11/5 = 0.160$

    a. In the x and y directions:

$$\left(S_y\right)_{cTc} = \sigma_o\left(\varepsilon_{qus}/(1+0.5\varepsilon_{qus})\right)^m = 110(0.264/(1+0.132))^{0.62}$$
$$= 44.6 \text{ ksi}$$

In the z direction:

$$\left(S_y\right)_{cLt} = \sigma_o\left(\varepsilon_{quo}/(1+2\varepsilon_{quo})\right)^m = 110(0.16/(1+0.32))^{0.62}$$
$$= 29.7 \text{ ksi}$$

    b. ε_{qus} and $\varepsilon_{quo} < \varepsilon_u$
In x and y direction:

$$\left(S_u\right)_{tTc} = \left(S_u\right)e^{\varepsilon_{quo}} = 44 e^{0.16} = 51.6 \text{ ksi}$$

In the z direction:

$$\left(S_u\right)_{tLt} = \left(S_u\right)oe^{\varepsilon_{quo}} = 44 e^{0.264} = 57.3 \text{ ksi}$$

c. In the x-z plane: in the z direction

$$(S_y)_{tLt} = \sigma_o(\varepsilon_{qus}/(1+0.2\varepsilon_{qus}))^m = 110(0.264/(1+0.0528))^{0.62}$$
$$= 46.7 \text{ ksi}$$

In x direction:
$$(Sy)tTc = 0.90(Sy)cTc = 0.90 + 44.6 = 40.2 \text{ ksi}$$

at an angle of 60° to the z direction:
$$(S_y)_{t\theta} = (60/90)(S_y)_{tTc} + (30/90)(S_y)_{tLt}$$
$$= 2/3 \times 40.2 + 1/3 \times 46.7 = 42.4 \text{ ksi}$$

REFERENCES

1. Amstead, B. H., P.F. Ostwald and M.L. Begeman, <u>Manufacturing Processes,</u> John Wiley & Sons, New York 1988.
2. Datsko, J. <u>Material Properties and Manufacturing Processes,</u> John Wiley & Sons, New York 1967.
3. Datsko, J. and W.V. Mitchell, "Changes in Mechanical Properties in Metal-Forming Processes," <u>Jr. Mat'ls Engr. & Performance,</u> Vol. 2, No. 2 1993.
4. Border, M.P. and J. Datsko, "Strain Strengthening by Cyclic Deformation," Proc. 9th SECTAM, Vol. 9, 1978.
5. Eary, D.F. and E.A. Reed, <u>Techniques of Pressworking Sheet Metal,</u> Prentice-Hall, NJ 1974.
6. Datsko, J. and C.T. Yang, "Correlation of Bendability of Materials with Their Tensile Properties," Trans. ASME, Vol. 82, 1960.
7. Hoffman, O. and G. Sachs, <u>Theory of Plasticity,</u> McGraw-Hill, New York 1953.

280 Chapter 8

STUDY PROBLEMS

8-1. An annealed nickel alloy has the following mechanical properties:
$\sigma_o = 230$ ksi, $S_y = 40$ ksi, $\varepsilon_u = 0.53$, $\varepsilon_f = 1.38$, $E = 28 \times 10^3$ ksi.
A 3" diameter bar is first stretched to a 2.800" diameter. It is then compressed to a 3.100" diameter. It is next stretched to a 2.900" diameter, followed by a stretch to 2.780." The last step is compression to a 3.000" diameter.

   a. Neglecting friction, what is the final load on the cylinder?
   b. A transverse tensile specimen having a 0.357" diameter × 1 1/16" gage length as machined from the cylinder after the deformation indicated in Problem 1. What should the yield strength and tensile strength values be according to the strain-strengthening relationships?

8-2. A 4 1/2" diameter bar of annealed 70%Cu - 30%Zn brass is cold drawn to 4" diameter. An 8" long cylinder is cut from the bar and compressed to a 6.4" length. The properties of the annealed brass in the longitudinal direction are: $S_y = 10$ ksi, $\sigma_o = 100$ ksi, $m = 0.55$, $\varepsilon_f = 1.50$. The transverse properties are the same except for ε_f, which is 1.20.

   a. What is the numerical value of the axial (longitudinal) tensile ultimate strength?
   b. What is the value of the transverse tensile ultimate strength?
   c. What is the numerical value of the longitudinal tensile yield strength?
   d. What is the value of the transverse tensile yield strength?

8-3. A heat of copper in the form of 1" diameter bars was tested for the effects of cyclic deformation. After annealing at 1250°F for 1 hr, 1/2" diameter tensile specimens gave the following typical results:
$(S_y)_o = 5$ ksi, $(S_u)_o = 31$ ksi, $\sigma_o = 65$ ksi, $\varepsilon_u = 0.36$, $\varepsilon_f = 1.25$.

In one test, a 0.488" diameter by 1 1/4" gage length specimen was axially loaded to give the following axial strains ε_w and associated final stress σ_w:

Deformation Strengthening and Formability 281

$\varepsilon_{w1} = -0.225$	$\sigma_{w1} = 35.9$	$\varepsilon_{w7} = -0.219$	$\sigma_{w7} = 44.5$
$\varepsilon_{w2} = +0.177$	$\sigma_{w2} = 38.3$	$\varepsilon_{w8} = +0.183$	$\sigma_{w8} = 46.3$
$\varepsilon_{w3} = -0.177$	$\sigma_{w3} = 40.9$	$\varepsilon_{w9} = -0.180$	$\sigma_{w9} = 46.3$
$\varepsilon_{w4} = +0.177$	$\sigma_{w4} = 43$	$\varepsilon_{w10} = +0.177$	$\sigma_{w10} = 47$
$\varepsilon_{w5} = -0.180$	$\sigma_{w5} = 43.6$	$\varepsilon_{w11} = -0.177$	$\sigma_{w11} = 47.2$
$\varepsilon_{w6} = +0.177$	$\sigma_{w6} = 44.7$	$\varepsilon_{w12} = +0.727$	$\sigma_{w12} = 60$

    a. Calculate the yield load and the maximum load <u>during</u> the sixth step (cycle) of deformation.
    b. Calculate the tensile yield strength of the material <u>after</u> the sixth step.
    c. Calculate the yield load and the maximum load <u>during</u> the seventh step (cycle) of deformation.

8-4. A heat of ETP copper in the form of 3" square bars was tested to study the effects of large cyclic ulastic strains. In the annealed (1250°F for 1 hour) condition, the material had the following tensile properties in the axial (z) direction of the bar:
$S_y = 5$ ksi, $S_u = 30$ ksi, $\sigma_o = 64$ ksi, $\varepsilon_f = 1.4$. The properties in the transverse (x and y) direction were the same except for ε_f which was 0.9.

Three-in. cubes were machined from the bar and annealed. Then they were sequentially compressed with good lubrication between hard, flat, smooth steel plates in the z, x, and y directions. When the magnitude of the strain in each of the three directions was equal, then the final shape was again a 3" cube.

Cube #1 was compressed 10% first in the z direction, then 10% in the x direction, and finally 10% in the y direction. Thus
$\varepsilon_{w1} = \varepsilon_z = -0.105;$ $\varepsilon_{w2} = \varepsilon_x = -0.105;$ $\varepsilon_{w3} = \varepsilon_y = -0.105$.

Cube #5 was given the following deformation:
$\varepsilon_{w1} = \varepsilon_z = -0.163;$ $\varepsilon_{w2} = \varepsilon_x = -0.163;$ $\varepsilon_{w3} = \varepsilon_y = -0.163$.

Cube #6 was given two complete cycles of 10% compression. That is
$\varepsilon_{w1} = \varepsilon_z = -0.105;$ $\varepsilon_{w2} = \varepsilon_x = -0.105;$ $\varepsilon_{w3} = \varepsilon_y = -0.105;$
$\varepsilon_{w4} = \varepsilon_z = -0.105;$ $\varepsilon_{w5} = \varepsilon_x = -0.105;$ $\varepsilon_{w6} = \varepsilon_y = -0.105$

282   Chapter 8

Calculate the tensile yield and tensile ultimate strengths in the three directions of the three deformed cubes.

8-5. A channel is made by drawing the two legs over a 1/8" die radius. (See Figure 8.2) The material used is 1/8" thick aluminum sheet having

$\sigma_o = 25$ ksi, $m = 0.25$, $S_y = 4.8$ ksi, $A_r = 85\%$, $E = 10 \times 10^3$

  a. What is the longitudinal tensile yield strength on the outside surface of the leg?
  b. What is the longitudinal tensile yield strength at the mid-thickness of the leg?
  c. What is the longitudinal tensile strength at the mid-thickness of the leg?
  d. What is the longitudinal tensile strength in the web of the channel?
  e. What is the longitudinal tensile strength at the outside surface of the leg?
  f. Plot the longitudinal tensile yield strength vs. the thickness h of the leg.

8-6. A 1" thick plate of stainless steel is bent to a 1/2" radius and is then unbent. The neutral axis remains at the mid-thickness.

$S_y = 40$ ksi, $\sigma_o = 230$ ksi, $\varepsilon_u = 0.53$, $\varepsilon_f = 1.38$, $E = 28 \times 10^3$ ksi

What is the value of the longitudinal tensile yield strength at a distance of 0.366" above the bottom surface of the plate. Assume that during bending, the bottom surface was the inside surface or fiber.

8-7. A 1/2" dia. × 2" long cylinder of annealed aluminum having the properties:

$S_y = 4.5$ ksi, $S_u = 12$ ksi, $m = 0.25$, $\varepsilon_f = 2.3$, $E = 10^4$ ksi

is twisted to fracture.

  a. How many degrees of rotation are needed?
  b. What is the approximate longitudinal tensile yield strength at the surface if it twisted 1° less than in Part a?
  c. At the center of the cylinder?
  d. What is the approximate longitudinal compressive yield strength at the surface?
  e. What is the final length of the cylinder?

Deformation Strengthening and Formability   283

8-8. A 2" diameter × 8" long cylinder of annealed 304 stainless steel is twisted 3 turns clockwise and then 3 turns counter clockwise. The original mechanical properties are isotropic and equal to:
$S_y = 35$ ksi, $S_u = 82.5$ ksi, $m = 0.45$, $\varepsilon_f = 1.67$

What is the maximum value of tensile yield strength at the surface, and at what angle to the axis of the cylinder does it occur?

8-9. To what angle ϕ can the cube of side length a be distorted to by means of direct or pure shear deformation? The thickness of the cube (not shown) remains constant at a. The cube is made of brass with the following properties:
$S_y = 10$ ksi, $S_u = 45.8$ ksi, $m = 0.44$, $\varepsilon_f = 0.76$, $E = 16 \times 10^3$ ksi.

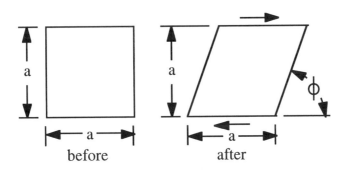
before   after

8-10. A cube ABCD of side length "1" is deformed in "pure shear" to an angle $\phi = 20°$. What are the magnitude and final direction of the maximum compressive strain?

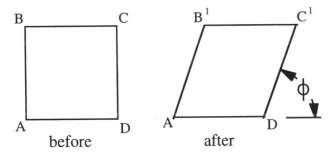
before   after

Chapter 8

8-11. A disc of diameter $D_D = 20"$ and thickness $h = 1"$ is formed into a cone of outside diameter $D_C = 20"$ and a diameter at the base $D_B = 4"$ by a shear-spinning operation. The half-angle ϕ of the cone is 30°. The material is a nickel base superalloy having the original properties:
$S_y = 45$ ksi, $S_u = 69.7$ ksi, $\sigma_o = 150$ ksi, $\varepsilon_u = 0.4$, $\varepsilon_f = 1.2$

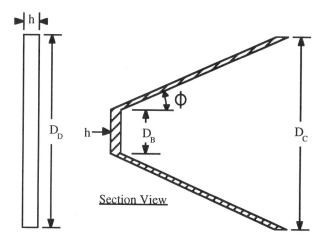

Section View

a. What is the numerical value of the maximum induced tensile strain? What is its direction?
b. What is the numerical value of the maximum compressive yield strength? What is its direction?
c. What is the numerical value of the minimum compressive yield strength? What is its direction?

8-12. An automobile manufacturer is considering the replacement of the steel guard rail in the door panel with an aluminum one. The steel is 1/8" AISI 1008 having the following properties along the length of the hat section after fabrication:

$(S_y)_w = 45$ ksi; $(S_u)_w = 50$ ksi; $A_r = 30\%$; $S_e = 25$ ksi

5456–0 aluminum is the replacement material. It has the following mechanical properties:

$(S_y)_o = 23$ ksi; $(S_u)_o = 45$ ksi; $\varepsilon_u = 0.23$; $\varepsilon_f = 0.8$

One possible method is to purchase 0.139" thick (non-standard) 5456–0 aluminum, cold-roll it to 0.125," and then deep-draw it into the hat section shape. A second possible method is to purchase standard 1/8" thick 5456–0 aluminum and do the deep drawing in two stages: first bend the flanges and web to some radius R_x, and then unbend them to make the finished hat section shown in the sketch.

What radius R_x is needed to give the finished hat section a tensile yield strength in the longitudinal direction at the surface of the web and flanges equal to 45 ksi?

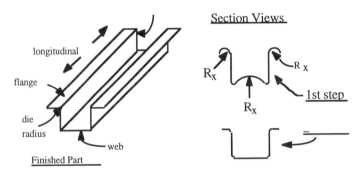

8-13. An annealed coil of 1/4" thick × 48" wide 304 stainless steel is cold rolled to 1/8" thickness. The 1/8" thick material is then sheared perpendicular to the rolling direction to 6" wide strips × 48" long. These strips are then drawn over a 1/16" die radius to form a 48" long hat section (see sketches below).
$S_y = 85$ ksi; $\sigma_o = 185$ ksi; $\varepsilon_u = 0.45$; $\varepsilon_f = 1.67$

   a. What is the compressive yield strength in the z direction at the outside surface of bottom B?
   b. What is the value of the tensile yield strength in the Y direction at the inside surface of the legs L?

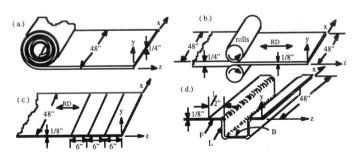

286 Chapter 8

8-14. A 1/8" thick by 8" dia. disc of aluminum is drawn over a 1/16" die radius to a 4" dia. cup. The original properties are:
$S_y = 5$ ksi; $\sigma_o = 23$ ksi; $m = 0.25$; $\varepsilon_r = 2.2$

   a. What is the magnitude of the circumferential strain at the top (open-end) of the cup just before it goes over the die radius?
   b. What is the thickness of wall at the corner tangency point near the bottom of the cup?
   c. List the sequence, including the magnitude and sense, of the longitudinal strains in the cup and radial strains in the disc at the inside surface near the top of the cup.

8-15. Calculate the force to draw a 1" dia. bar to 0.900" dia. (neglect friction and redundant work). The material is annealed 1100-0 aluminum having the properties:
$E = 10 \times 10^3$ ksi; $S_y = 5$ ksi; $S_u = 12.1$ ksi; $\sigma_o = 22$ ksi;
$m = \varepsilon_u = 0.25$; $\varepsilon_f = 2.3$

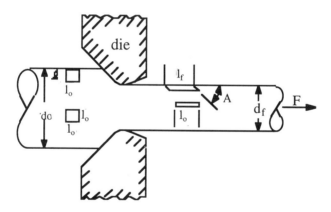

8-16. What is minimum dia. to which the 1" dia. bar of problem 8-15. could be drawn to in one step? Neglect friction and redundant work.

Chapter 9

STRENGTH OF WELDMENTS

INTRODUCTION

This chapter presents a new approach to the study of the load-carrying capacity of a weld joint. It is a quantitative method that permits the design and production of weldments that are more efficient in terms of cost, weight, and reliability. This new method is based on the simple concept that the load that any member can carry is the product of the strength of the material and some function of the geometry of the part. For simple axial loads, the only geometric factor needed is the cross-sectional area. For parts subjected to bending, the moment of inertia and length must be known.

The method of strength analysis presented in this chapter is based upon the microstructure and hardenability principles presented in preceding chapters. The strength is predicted from the microstructures that are present after welding; the microstructures are predicted from hardenability data and cooling rates.

In regard to the geometry of the weldment, emphasis is placed on relationships between the welding variables and the two important size factors: (1) the size of the bead or fillet and (2) the depth of penetration (fusion). While the discussion is for the arc processes (stick, mig, submerged, etc.) and the welding of steel, similar treatments can be used for the other processes and materials.

THE WELDING PROCESS

Before studying the strength of welded joints it is best to understand what welding is. The author's definition is geared to the engineer. *Welding is the manufacturing process of joining two or more pieces of solid material together by utilizing only the four fundamental attracting forces that hold the atoms of a solid together.* These bonding forces are discussed in Chapter 2 and are simply listed here in order of decreasing bonding strengths: (1) covalent, (2) ionic, (3) metallic, and (4) molecular. Thus the joining of two pieces of plastic or the joining of pieces of metal or ceramic by means of an adhesive is truly welding since the forces holding the pieces together are van der Waals' molecular attractive forces. Welding is not restricted to the joining of metals by melting or fusion. In the cold-pressure welding of metals, no heat is utilized. The weld is achieved by severe plastic deformation at the interface. The fundamental welding problem is the removal of the adsorbed gas and oxide layers from the intended contacting surfaces so that the original surfaces are separated by only the grain boundaries of the metal.

TYPES OF WELDING PROCESSES

As outlined in Table 9-1, there are four types of welding processes, in terms of the energy used to remove the gas and oxide layers from the contacting surfaces. The first type uses only mechanical means, or severe plastic deformation at the interface, to achieve a weld. Although it has been used industrially as a welding process since about 1950, it is not used as extensively as it could be. Its main application today is in the joining of the ends of aluminum and copper coils of wire used in cold heading machines, in the joining of electrical lead wires to their terminals, and in encapsulating electronic devices. Since the process entails severe plastic deformations, it is used mainly with ductile materials such as aluminum and copper. This process is illustrated in Figure 9-1.

The second type, mechanical plus thermal, contains both the oldest and the newest of the welding processes. Undoubtedly the oldest welding process is the forge welding process, where the pieces to be joined have been heated in an open fire or "forge," and then the ends to be joined are placed, one on top of the other, on an anvil and and hammered (or forged) together. Early practitioners of this trade developed considerable skill in the joining of bars or plates of iron. Because of the limitations in the size of castings and forgings that could be produced, whenever large parts were required, it was necessary to employ the

Strength of Weldments   289

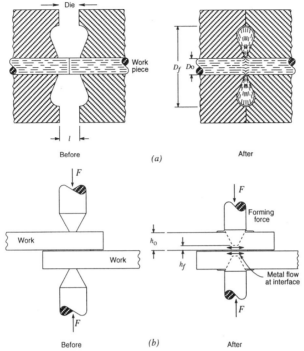

Figure 9-1. Schematic drawing of typical cold-welding operations: (a) butt weld of bars and (b) lap-spot weld of sheets.

forge welding process. Welding artisans achieved surprisingly good welds by first crowning or beveling the mating surfaces to facilitate squeezing out the oxides. By throwing silica sand onto the areas to be welded, they converted the iron oxide to a more fluid slag, which aided its expulsion from the interface.

Recrystallization welding and friction welding are very recent innovations. The former has limited applications because it lies between the cold and the hot processes and does not have the advantages of either. Because of its simplicity, friction welding is becoming more widely applied as an industrial welding process. In friction welding, one part is clamped in a machine tool resembling a lathe, and then its end is rotated while lightly contacting the mating part (which is also clamped). After a short time, when the

Table 9-1. Welding Types Based on the Energy Used to Remove the Adsorbed Gas and Oxide Layers in Order to Get Metal-to-Metal Contact

1. Mechanical
    a. Cold pressure in inert atmosphere
    b. Cold pressure in a vacuum
    c. Cold pressure – severe deformation in air
2. Mechanical plus thermal
    a. Warm pressure: recrystallization
    b. Hot pressure: forge; friction
3. Thermal
    a. Gaseous fuel: oxyacetylene; propane
    b. Solid fuel: thermit
    c. Electric: resistance; induction; arc; electron beam
    d. Laser
    e. Warm air: for welding plastics
4. Chemical plus thermal: heated flux to remove surface films

interface approaches the melting temperature, the two parts are pressed together with a large force as the relative motion is stopped.

During the first half-century most industrial welding has been thermal, type 3, in Table 9-1. The most common gaseous fuel, a combination of compressed oxygen and acetylene, is used because of its relatively high flame temperature, which is in excess of 5,000°F. This welding process is illustrated in Figure 9-2.

Originally the d–c electric arc was by far the most common of the electric heat sources, but now the a–c arc and the resistance processes are also widely used. The temperature of the arc column is approximately 12,000°F, and the temperature at the electrode tip is approximately 18,000°F. When arc-welding steel, the cooling rate may vary from a low value of about 10°F/sec to a high value of about 100°F/sec at 1,300°F. However, the typical values are in the range of about 30 to 60°F/sec. Under these conditions, the usual microconstituents for plain carbon steel are ferrite and fine pearlite. The arc-welding processes are illustrated in Figure 9-3.

The common arc processes (manual or stick, MIG, TIG, submerged arc) utilize voltages in the range of 10 to 40, and currents in the range of 50 to 1200 amperes. The electron beam process, on the other hand, utilizes voltages in the kilovolt range and currents in the milliamp range.

Strength of Weldments 291

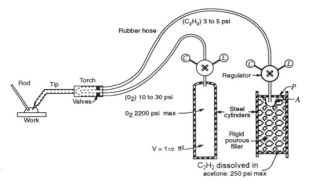

Figure 9-2. Typical gas welding setup.

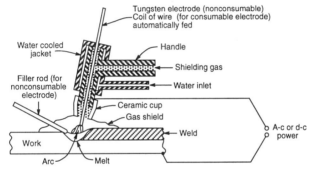

Figure 9-3. Inert gas-shielded, arc-welding process (IGSA). The separate filler rod is not used when the wire-feed, consumable-electrode process is used.

The electric resistance welding process is illustrated in Figure 9-4. Only four to six cycles of 60 cycle current are needed to achieve a resistance spot weld in steel with a potential of only three to four volts but a current of several hundred amperes. The metal at the joint interface is heated to the melting temperature in about 0.1 seconds, and the cooling time is not much longer. Therefore, cooling rates of 1,000°F/sec at 1300°F are possible, with the resulting microconstitutent being the brittle but hard martensite. As an aid to the prevention of the formation of martensite, most modern spot-welding equipment includes a programmable controller that makes it possible to provide preheating and postheating treatments of several cycles of low current.

292 Chapter 9

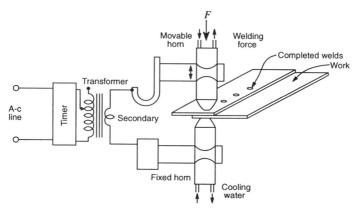

Figure 9-4. Electric-resistance spot-welding process.

In laser welding, a concentrated light beam is focused on the region to be welded. The high-intensity heat source makes it possible to have a high welding velocity. In addition to being an industrial manufacturing process, laser beam welding has medical applications.

In some cases, the melting temperature of the oxide is higher than that of either the parent metal or the filler material. Then a flux must be used that will combine chemically with the oxide layer to form a low-melting-temperature slag that can be displaced easily from the weld interface. This fourth type of welding is the chemical plus thermal process. In the past, it would be associated with the brazing and soldering processes.

CALCULATION OF THE SIZE OF THE WELD

There are two components that make up the size of the weld nugget: the bead or fillet, and the penetration. The bead or fillet comes from the electrode or filler rod, and the metal in the penetration region is part of the original plate or bar. To calculate the strength of the weld nugget, it is assumed that there is no mixing of the two regions. That is, all of the metal in the bead is the same composition of the welding rod and all of the metal in the penetration is the same composition of the original plate. When two plates of different composition are welded together, it is assumed that the composition of the metal in the penetration zone is the average of the two.

Burn-Off Rate (BOR)

To determine the size of bead or fillet from the welding conditions, it is necessary to know how the arc voltage and current affect the rate at which the electrode melts. Experimental work done in the author's laboratory with the E6XXX series of electrodes as well as with the submerged arc indicates that the BOR can be estimated with reasonable accuracy from the equation:

$$Q_m = 4.3 \times 10^{-5} EI \tag{9-1}$$

where Q_m is the cubic inches melted per arc minute, E is the arc voltage, and I is the arc current.

Not all of the metal melted is deposited as the bead or fillet. For some electrodes such as the E6010 and E6012 type, 10 to 20% of the metal of the electrode (depending upon the size of the electrode) is wasted as "spatter loss," that is, small particles that bounce onto the shop floor or onto the plate being welded but away from the weld joint. Some electrodes have some iron powder in the coating that ends up in the weld bead. The E6014 type has enough iron in the coating to offset the metal lost by spatter so that the net spatter loss is zero. Other electrodes, such as the heavily coated 7024 type actually have a negative spatter loss of 20% or more.

The metal deposited can then be estimated from the equation:

$$Q_d = Q_m(1 - SL) \tag{9-2}$$

where Q_d is the cubic inches deposited per arc minute and SL is the spatter loss expressed as a fraction rather than a percentage.

Bead Size

The cross-sectional area of the bead can be calculated from the BOR and the welding velocity (see Figure 9-5). By making a mass balance, that is equating the metal deposited per arc minute to the product of the area of the bead and the welding velocity, we get the relationship:

$$Q_d = A_b \times v \tag{9-3}$$

294   Chapter 9

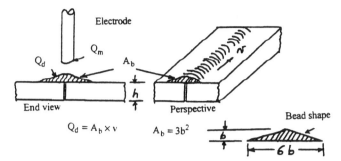

Figure 9-5. Weld bead that has a good shape.

where A_b is the cross-sectional area of the bead. From this we get:

$$A_b = Q_d / v \qquad (9\text{-}4)$$

In order to calculate the load carrying capacity of the bead it is necessary to know the shape of bead, particularly its height. A good assumption in regard to the shape of a well-proportioned bead is an isosceles triangle having a base that is six times the height, as illustrated in Figure 9-5. If the height of the bead is b, then $A_b = 3b^2$ and

$$b = (A_b / 3)^{0.5} \qquad (9\text{-}5)$$

Since $\quad A_b = Q_d / v$

then $\quad b = (Q_d / 3v)^{0.5} \qquad (9\text{-}6)$

The height of the bead now can be expressed in terms of all of the welding variables as:

$$b = \left[4.3 \times 10^{-5} EI(1 - SL) / 3v\right]^{0.5} \qquad (9\text{-}7)$$

It must be noted that this relationship is valid only for steel.

Depth of Penetration

For arc-welding steel plates in a square groove butt-joint with no gap between the plates and no preheat, the following two equations give a good indication of the depth of penetration.

$$p = 2 \times 10^{-5} E^{0.56} I^{1.40} v^{-0.29} \quad \text{for} \quad I \leq 200 \text{ amp} \quad (9\text{-}8)$$

$$p = 2 \times 10^{-5} E^{0.56} I^{1.33} v^{-0.33} \quad \text{for} \quad I > 200 \text{ amp} \quad (9\text{-}9)$$

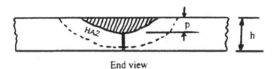

End view

Figure 9-6. Penetration of a butt joint.

These equations are valid for p less than h/2, where h is the plate thickness, as shown in Figure 9-6. If p is greater than h/2, the calculated value will be less than the actual penetration.

AXIAL LOAD CAPACITY OF BUTT JOINTS

The axial load that a butt joint can carry is equal to the sum of the load that the bead can carry plus the load that the penetration zone can carry, as shown in Figure 9-7. As stated earlier, the load is equal to the product of the area and the strength. It is apparent from this figure that the heat-affected zone (HAZ) carries no load. Thus:

$$L_w = S_b \times b + S_p \times p \qquad (9\text{-}10)$$

where S_b and S_p refer to the strength of the bead and penetration, respectively.

The yield strengths are used to calculate the yield load, and the tensile ultimate strengths are used to calculate the load that will cause complete failure. When calculating the cyclic load that will cause fatigue failure, the appropriate

296   Chapter 9

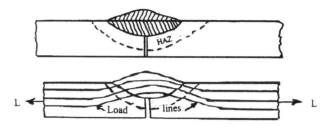

Figure 9-7. Axial load capacity of a butt joint.

stress concentration factor for the notch at the base of the penetration must be used.

If equal welds are made on both sides of the plates, then the load calculated for one weld is multiplied by two.

In designing an arc-welded butt joint, a good criterion to begin with is a weld joint that has a load-carrying capacity equal to that of the plate. Such a weld can be described as having a joint efficiency of 100%. If a 20% factor of safety is desired, then the joint efficiency would be 120%.

A relationship between the welding variables and the load capacity can be developed for any desired joint efficiency. Such a relationship for 100% joint efficiency, and a weld on only one side of the plates is presented below. The load capacity of the plate is

$$L_o = S_o \times h \tag{9-11}$$

where S_o is the appropriate strength of the plate and h is the plate thickness. The load capacity of the weld joint can be expressed in terms of the welding variables and equated to L_o as shown in Eq. (9-12).

$$S_o \times h = S_b \left(4.3 \times 10^{-5} EI(1-SL)/3v\right)^{0.5} + S_p \left(2 \times 10^{-5} E^a I^b v^{-c}\right) \tag{9-12}$$

The three welding variables in this equation are E, I, and v. E can vary only a small amount since it is influenced by the arc length. The two important variables are I and v. If the welding cost is to be minimized, then I and v should be as large as practical and joint efficiencies much greater than about 120% should be avoided.

The important aspect of the above equation is that it permits both the design engineer and the welding engineer to determine the optimal welding speed for any current and load capacity.

Strength and Microstructures

The methods of predicting the strength of steel from its microstructure and methods to predict the microstructure from hardenability data and cooling rates were presented in earlier chapters. To predict the strength of the steel in a weld joint requires some means to determine what the cooling rate is for a given set of welding conditions or heat input. Such calculations can be made by anyone knowledgeable about heat transfer, but they may require some experimentally determined constants.

To simplify the problem of determining the cooling rate after arc welding, the author had the nomograph shown in Figure 9-8 constructed. It is based on a line heat source moving along the edge between two plates in a square groove butt joint. The equivalent energy input is the product of the arc wattage (EI) and an efficiency factor (f) that varies between 0.80 for a long, exposed arc column to 1.0 for submerged arcs, divided by the plate thickness. Thus

$$EEI = EIf/h \qquad (9\text{-}13)$$

To use the nomograph one first determines the equivalent energy input and intersects it with the welding velocity. Then a vertical line is drawn to intersect with the appropriate horizontal preheat temperature line. Finally a line is traced downward and to the left at 45° until it intersects the cooling rate scale at the left margin. The resulting value is the cooling rate in °F/sec at 1300°F that can be used in conjunction with the previously presented hardenability data to determine the hardness, microstructure, and strength of the steel.

For example, consider the case of arc welding 1/4" thick plates of low carbon steel with 25 volts, 150 amperes, 10 ipm welding velocity and an arc efficiency of 1.0. The equivalent energy input (EEI) is $25 \times 150 / (1/4) = 15 \text{ kw/in}$. By locating the intersection of the 15 kw/in. line and the 10 ipm line and moving vertically in Figure 9-8 to the 60°F initial temperature line and then following the 45° lines to intersect the cooling rate axis, one obtains a value of 40°F/sec at 1300°F.

298   Chapter 9

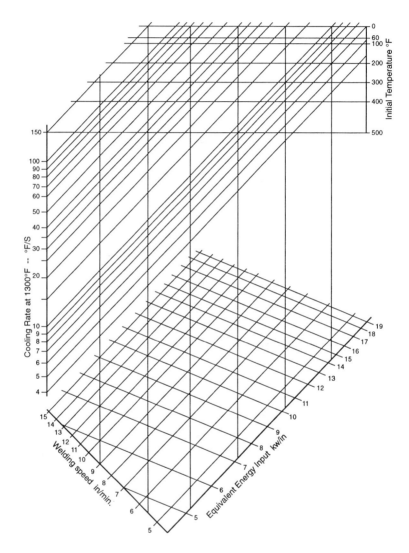

Figure 9-8. Nomograph for the cooling rate at 1300°F during arc welding of steel.

The following equation[1] can also be used to estimate the cooling rate when welding thin plates — plates thin enough so that the temperature difference between the bottom and top surfaces is small in comparison to the melting temperature.

$$R = 2\pi k\rho C (h/H)^2 (T_c - T_o)^3 \qquad (9\text{-}14)$$

R is the cooling rate °C/s at temperature T_c which should be taken as 704°C (1300°F). k is the thermal conductivity $(J/mm \cdot s \cdot °C)$ and ρC is the volumetric specific heat $(J/mm^3 \cdot °C)$. h is the plate thickness (mm) and T_o is the original plate temperature.

H (J/mm) is the net heat input and is equal to $f \cdot E \cdot I / v$, where f is arc efficiency factor, E and I are the voltage and current, and v is the welding velocity in mm/s.

The average value of k can be taken as 0.035, and ρC as 0.006 for the range of temperatures encountered in arc-welding steel. Then the equation for the cooling rate of steel in the weld zone can be reduced to

$$R = 13 \times 10^{-4} (h/H)^2 (704 - T_o)^3 \; °C/s \qquad (9\text{-}15)$$

When the conditions listed in the above example are substituted in the above equation, the cooling rate is calculated as 21.6°C/s or 41°F/s, which is in good agreement with the value obtained from the nomagraph.

Table 9-2 illustrates the advantage of welding with the maximum allowable current and a correspondingly high welding velocity. When welding with low current and velocity (125A, 4 ipm, and 45 kw sec/in.), the penetration is only 0.05" and the HAZ is 0.109." Considering that the HAZ does not carry load, this is a poor combination of welding conditions.

On the other hand, with high values of current and velocity (800A, 27.4 ipm, and 45.5 kW sec/in.), the penetration is increased to 0.26" and the HAZ is reduced to 0.054." Not only is the penetration increased by a factor of 5, but the time to weld a given length of joint is reduced by a factor of 7.

---

[1] Jhaveri, P., W. G. Moffatt, and C. M. Adams, Jr. The effect of plate thickness and radiation on heat flow in welding and cutting. Welding Journal 41(1):1962; 12s-16s

Table 9-2. Efficiency of Heat Utilization in Arc Welding

Current, Amperes	Potential, Volts	Velocity, ipm	Total Energy Input		Penetration, in.	Nugget Width, in.	Area, sq. in.	HAZ Depth, in.
			Watt sec/in.	Btu/in.				
125	24	4.0	45,000	42.7	0.05	0.45	0.042	0.109
170	27	6.0	45,700	43.3	0.10	0.59	0.067	0.101
300	32	12.6	46,500	44.1	0.15	0.59	0.094	0.078
410	26	14.0	45,700	43.3	0.16	0.62	0.101	0.066
800	26	27.4	45,500	43.2	0.26	0.56	0.140	0.054

FILLET WELDS

The size and load capacity of fillet welds can be analyzed in much the same manner as previously used for the butt joint.

Fillet Size

The BOR for fillet welds is the same as for the butt welds except that in some cases different electrodes are used. The area of the fillet, A_f, is $1/2 \ f^2$ for the fillet shape shown in Figure 9-9, where the fillet size is designated as f.

The fillet size can be expressed as a function of the welding variables by making a mass balance.

Thus
$$Q_d = A_f \times v = 0.5 f^2 v \qquad (9\text{-}16)$$

Also
$$Q_d = 4.3 \times 10^{-5} \ EI(1 = SL) \qquad (9\text{-}17)$$

So
$$f = \left(2 \times 4.3 \times 10^{-5} \ EI(1-SL)/v\right)^{0.5} \qquad (9\text{-}18)$$

Strength of Weldments

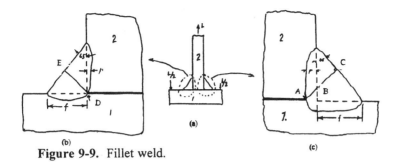

Figure 9-9. Fillet weld.

Load Capacity of Fillet Welds

The tensile load capacity of fillet welds can be determined by Figure 9-9. Figure 9.9(a) shows a TEE joint with two fillet welds joining plates 1 and 2 subject to a tensile load L.

Figure 9-9(b) is an enlargement of a fillet made when a low welding current (I < 200A) is used. Although there is penetration in both plates 1 and 2 near the middle of the fillet, there is no penetration or fusion at the root D of the weld joint. Thus all of the load is carried by the fillet. A good assumption is that failure will occur by shear along the line DE. Thus the load per inch length of weld can be estimated by multiplying the length DE by the shear strength of the metal. That is

$$L_w = DE \times (S_s)f \qquad (9\text{-}19)$$

The shear strength can be assumed to be equal to one-half the tensile strength: $(S_s)f = (S_t)f/2$. Also, DE can be expressed in terms of the fillet size f: $DE = f/2^{0.5}$. Thus

$$L_w = 0.707 \left(2 \times 4.3 \times 10^{-5} \, EI(1-SL)/v\right)^{0.5} \times 0.5(S_t)f \qquad (9\text{-}20)$$

An enlargement of the weld zone for fillets made with large currents (I > 200A), is shown in (c). Here there is penetration into the root of plate 2, an amount equal to the depth of penetration p. To calculate the penetration p, use

302   Chapter 9

I/2 for the current since half the current goes into plate 1 and half into plate 2. Now failure occurs by shear along the line ABC. If the filler material from the electrode is of different composition than the plates, then the strength of the metal in the penetration zone is different from that of the metal in the fillet. Therefore:

$$L_w = AB \times (S_s)p + BC \times (S_s)f \qquad (9\text{-}21)$$

where L_w is the load per inch length of weld. But AB = 1.414p and BC = 0.707(f − p). Thus the load can be expressed as:

$$L_w = 1.414p \times 0.5(S_t)p + 0.707(f-p) \times 0.5(S_t)f \qquad (9\text{-}22)$$

As shown previously, both p and f can be expressed in terms of the three important welding variables, E, I, and v.

A fillet weld such as that shown in Figure 9-9 can be subjected to other types of loading, such as bending. The method of determining the size of the weld needed in this case is the same as the one presented above. What is different is the calculation of the stresses in the weld joint due to the load. This latter problem is one of applied mechanics and is well treated in mechanics textbooks.

EXAMPLE PROBLEMS

9-1.   1/2" thick annealed 1330 steel plates are arc-welded in a square groove butt joint. The electrode is 1010 steel wire and the welding process is submerged arc. The welding conditions are: E = 35 v, I = 350 A, v = 15 ipm, and no preheat. The composition of the 1330 steel is: 0.30% C; 1.7% Mn; 0.25% Si; and a #5 G.S. What is the probable hardness at the upper HAZ (near the region where melting occurred)?

Strength of Weldments

Solution:
v = 15 ipm = 6.35 mm/s
H = (35×350)/6.35 = 1929 J/mm
R = 13×10^{-4}(12.7/1929)^2(704−20)^3 = 18°C/s = 32°F/s

This is equivalent to 8 Jominy distances. From Tables 6-8 and 6-9 we get:

f_C = 0.32×0.3+0.123 = 0.22
f_{Mn} = 5.125×1.7−1 = 7.71
f_{Si} = 0.7×0.25+1 = 1.18

The DI = 1.99 and the dividing factor f_d = 1.78
R_c of primary martensite: R_c = 73+17 ln 0.3 = 52.5
R_c at HAZ = 52.5/1.78 = 29.5, or 280 H_B

9-2. Is any martensite present in the steel HAZ of Problem 9-1 if the hardness is 280 H_B?

Solution:
From Chapter 4:
C_E = 0.80 − 0.061×1.7 − 0.125×0.25 = 0.67
If no martensite is present, then the microconstituents are ferrite and pearlite
f_p = 0.3/0.67 = 0.45 and f_α = 0.55
assume fine pearlite
H_B = 0.45×380 + 0.55×80 = 215
Yes, some martensite is present.
Then 280 = 0.55×80 + (y×741) + (0.45−y)×380
where y = fraction of martensite with 0.67 C and y = 0.18, or 18%.

304   Chapter 9

REFERENCES

1. AWS, <u>Welding Handbook</u>. The American Welding Society, Section 2, NY 1984.
2. Blodgett, O.W. <u>Design of Weldments</u>. Lincoln Arc Welding Foundation, OH 1965.
3. Datsko, J. <u>Material Properties and Manufacturing Processes</u>. John Wiley & Sons, NY 1967.
4. Cary, H.B. <u>Modern Welding Technology</u>. Prentice-Hall, NJ 1979.

STUDY PROBLEMS

9-1. 1/4" thick plates of 1020 annealed steel are arc-welded with 0.2 C carbon electrodes. The joint is a square groove butt weld with welding done on 2 sides. The welding conditions are 30 V, 150 A, 10 ipm, and a cooling rate of 50°F/sec. What load per-inch weld will the weldment carry? Assume 0% spatter loss.

9-2. A T-weld is made on 1/4" thick 1020 steel (annealed), as sketched. The filler material is also 1020 steel. The welding conditions are: E = 30 V, I = 300 A, v = 10 ipm, S = 0%, and CR = 40°F/sec at 1300°F. Calculate Lu. Assume 0 spatter loss.

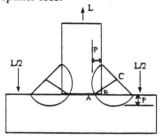

Strength of Weldments 305

9-3. 1/4" thick plates USS Cor-Ten steel (A242) are arc-welded in a square groove butt joint. The catalog gives the following data for this steel. Typical composition: 0.16 C, 1.15 Mn, 0.20 Si, 0.32 Cu, 0.50 Cr, 0.05 V. As-received mechanical properties: min. $S_y = 50$ ksi; min. $S_u = 70$ ksi; min. 2" El = 24%; min. 8" El = 19%. The eutectoid composition is approximately 0.65 C. The hardenability band values are as follows: 1/8"= 44-33; 1/4"= 38-24; 3/8"= 31 max.; 1/2"= 27 max.; 4/8"= 25 max.; 3/4"= 23 max.; 7/8"= 22 max.; 1"= 21 max.; 1-1/4"= 20 max. The electrodes used are E7018, 5/32" diameter having the characteristics $E = 321 + 25$. The composition is: 0.06 C; 1.10 Mn; 0.50 Si. The as-welded mechanical properties are: $S_y = 68$ ksi; $S_u = 70$ ksi; 2" El = 31%. The welding is done with a 5/32" arc length, 175 A, 13 ipm, 80% arc efficiency.

    a. If a good shape results, what is the height of the bead?
    b. What is the depth of penetration?
    c. On the basis of the maximum values of the hardenability band, what microstructures and approximately how much of each will be present?
    d. What axial tensile load per inch of weld can the weldment carry? Neglect eccentric loading.

9-4. Two 1/4" diameter of annealed nickel wire are cold-pressure butt-welded. The diameter of the "upset" is 0.450", and this "flash" is removed after welding. The relationship $\sigma = 80,000 \, \varepsilon^{0.3}$ describes the plastic flow characteristics for the annealed nickel. Its other mechanical properties are: $S_y = 12$ ksi; $S_u = 43.1$ ksi; $A_r = 80\%$; $E = 30 \times 10^3$ ksi. Assume that all of the oxide is dispersed from the original interface.

    a. What yield load will the wire outside the weld area carry?
    b. What yield load will the "weld interface" carry?
    c. What maximum tensile load will the wire outside the weld carry?
    d. What maximum tensile load will the "weld" carry?

Chapter 10

MACHINABILITY OF METALS

INTRODUCTION

Before discussing the material property or characteristic known as "machinability," it is desirable to discuss briefly the manufacturing process of machining.

The technological advances in the field of machining during the past 100 years have made this manufacturing process a very important and essential one. It may be said that the machine tool is to our mechanized era what power was to the industrial revolution. In fact, without machine tools, the engines that provide our industrial and domestic power would not exist. For example, although James Watt *designed* a steam engine with a separate condenser in 1775, it took 25 years more to *make* it — even with the help of one of England's best machine builders. In Watt's day, it was considered a major achievement when a cylinder was bored so that when a piston was tightly fitted at one end, a clearance not greater than the thickness of a shilling was present at the other end. Today, cylinders can be bored to an accuracy of a thousandth of an inch (0.001 in.) without much difficulty and, with care, to a ten-thousandth of an inch (0.0001 in.).

During the past 150 years, American manufacturing industries have led the rest of the world in terms of quantity and efficiency of production. This superiority stems from America's pioneering achievements in *mass production*, a manufacturing technique conceived in the United States in 1800 by Eli Whitney for the manufacture of muskets. Prior to that time muskets were custom-made, even in the government's armory at Springfield, Massachusetts; soft parts were fitted together by the "trying and filing" technique, then hardened and set. Consequently, no two muskets were exactly alike, and their parts could not be interchanged. At a time when France and England were at war and the United States was at risk of becoming involved, Whitney was able to tremendously impress a board of government officials and army officers who saw the first demonstration of the results of mass production. Out of a box containing ten dismantled muskets, parts were selected at random and, before the astonished eyes of the officials, ten complete muskets were rapidly and easily assembled. But even more amazing than this feat, the jigs and fixtures and machine tools of Whitney's factory could produce with ease thousands and thousands of muskets, all exactly alike. Without machine tools and the machining process, American industrialization would not have been possible.

DEFINITION OF MACHINING

Machining is the manufacturing process in which the size, shape, or surface properties of a part are changed by *removing* the excess material. This is in contrast to *forming*, where the material is simply *moved*. Machining may be accomplished by locally straining the material to fracture through the relative motion of a tool and the work piece, as is done in the conventional processes. The material removal mechanism may be chemical, electrical, or thermal as in the case of unconventional processes such as chem-milling, electrolytic maching, and electro-arc machining. In this light, flame cutting is a chemical machining process.

Machining is a relatively expensive process that should be specified only when high accuracy and good surface finish are required, a condition that, unfortunately, is necessary for almost all mating parts. Today machining is performed almost exclusively by means of *machine tools*, which are power-driven machines that have provisions to hold and move the work piece as well as a *cutting tool*. These terms will all be described briefly.

CUTTING CONDITIONS

The cutting conditions in machining usually refer to those variables that are easily changed at the machine tool by the operator and that affect the rate of metal removal. They are: the tool material and shape, the cutting speed, and the size of cut, which refers to the feed and depth. The tool materials and the effect of tool shape on the performance of the tool are discussed in following sections. In general, we can say that the keener, or sharper, the cutting edge, the better the performance of the tool.

The *cutting speed*, v, is defined as the largest of the relative velocities between the cutting tool and the work material, and is expressed with the units feet per minute. In some machines it is the work that reciprocates or rotates to provide the cutting speed, whereas in other machine tools it is the cutter that moves to provide the cutting speed. The effect of the cutting speed on the tool life (the number of minutes that a cutting tool can cut before the cutting edge is worn away) is an exponential relationship that may be expressed as $t = Kv^{-m}$, where t is the time in minutes of cutting time until the tool becomes dull, v is the cutting speed, and K and m are constants for a given cutting condition. The constant K is very large, on the order of 10^8, and the exponent m is on the order of 10. Thus, it is apparent that for a small decrease in the cutting speed, the tool life will be increased greatly.

In general, it is more convenient in expressing the relationship of cutting speed to tool life to violate the principles of dependent and independent variables and write it as $vt^n = C$. In this form, C is a constant for each cutting condition, with a range of values from 10 for very hard steel to 10^4 for very soft magnesium, when the machining is done with a high-speed steel tool. In general, for a given work material, the value of C for a carbide tool is three to five times as large as for an HSS tool; for a ceramic tool it is five to eight times as large. In reality, C is the cutting speed that results in a one-minute tool life. Thus a material having a C value of 150 will cause the tool to fail in one minute if it is machined at a cutting speed of 150 fpm.

The exponent n varies from 0.05 to 0.12 for most work materials when a high-speed steel tool is used, and whenever actual values are not known, the value of 0.07 may be used. The average value of n for carbide and ceramic tools is generally given as 0.2 and 0.4, respectively. This equation, known as the Taylor equation, is discussed in more detail in the section "Tool Life" (p. 311).

The *depth of cut* d is defined as the distance the cutting tool projects below the original surface of the work, and is expressed in thousandths of an inch. The depth of cut determines one of the linear dimensions of the cross-sectional area of the size of cut. In general, an increase in the depth of cut will result in a nonlinear increase in temperature and decrease in tool life.

The *feed* f is defined as the relative lateral movement between the tool and the work during a machining operation. On the lathe and drill press, it is expressed with the unit inches per revolution (ipr). On the shaper, it is expressed as inches per stroke (ips). On the milling machine, it is expressed in inches per tooth (ipt), although the machine controls are designed with the units

310   Chapter 10

inches per minute (ipm, which is the product of the basic feed × the number of teeth in the cutter × the revolutions per minute of the cutter). The feed is the second linear dimension that determines the cross-sectional area of the size of the cut. In determining the cutting time for a given operation, the feed FR (in./min.) is used. This is the product of the feed f and the rpm or spm (strokes per minute).

The product of the speed, feed, and depth of cut determines the rate of metal removal, which is expressed as cubic inches per minute. On the lathe and other turning operations, the rate of metal removal is obtained from the expression 12 vfd. In drilling, for example, it would be the product of the feed (ipr) × the RPM of the drill × the cross-sectional area of the drilled hole. The effect of the size of cut on the efficiency of machining is discussed in detail in the sections "Tool Life" and "Machinability."

CHIP-FORMING MACHINE TOOLS

There are five basic types of machine tools: lathe or turning machine; drilling or boring machine; shaper or planer; milling machine; and grinder. The first four, referred to as the basic machine tools, are similar in that they use cutting tools that are sharpened to a predetermined shape; whereas, the cutting edges on a grinding wheel are not controlled. In general, the accuracy and surface finish obtainable on these machine tools may be summarized as follows: (1) for the basic machine tools, a surface finish of 50 to 250 rms (microinches), and an accuracy of ± 0.005 in. with ease; (2) for the grinding machines, a surface finish of 5 to 60 rms, and an accuracy of ±0.0001 in. with care and ±0.001 in. with ease.

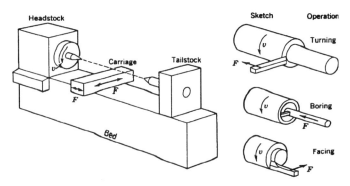

Figure 10-1. Lathe.

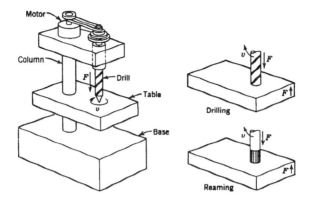

Figure 10-2. Drill press.

The basic motions possible with metal-cutting machine tools and the most commonly performed operations are illustrated in Figures 10-1 through 10-5. The surfaces that may be generated by machine tools fall into four general types: (1) planar, (2) external cylindrical, (3) internal cylindrical, and (4) irregular. In general, planar and irregular surfaces are machined on the shaper, milling machine, and lathe. External and internal cylindrical surfaces are created on lathes and drilling machines. Although drilling machines usually produce internal cylindrical surfaces, they are sometimes used to produce internal flat surfaces in operations such as spot-facing or counterboring. Grinding machines can produce any of the surfaces that the four basic machine tools can produce and with much better surface finishes and greater accuracy, but with correspondingly reduced rates of metal removal. Consequently, grinding is usually performed only as a final finishing operation.

In today's era of mass production and automation, many special-purpose machine tools combine operations of all four of the basic machines. Consequently they cannot be readily classified as any of the basic basic machine tools.

TOOL LIFE

The preferred method of describing the efficiency of machining is to state the rate of material removal — cubic inches of material removed per minute. However, since it is more convenient in the shop to evaluate cutting tools or work

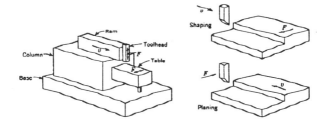

Figure 10-3. Shaper.

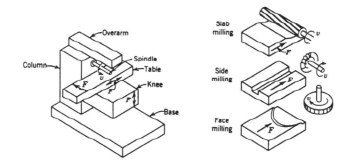

Figure 10-4. Milling machine.

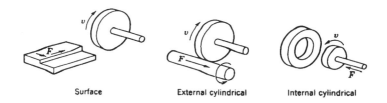

Figure 10-5. Operations performed on a grinding machine.

materials by means of their relative tool life under constant cutting conditions, the presently accepted way in industry of making this appraisal is by obtaining the values of n and C in the Taylor equation $vt^n = C$. Testing or data-collecting projects in machining are conducted most expediently by employing cutting speeds higher than those used commercially. These higher speeds result in a shortened tool life but offer a consequent saving in expensive testing time. Figure 10-6 illustrates the results of a test conducted by the author to check the validity of extrapolating the tool life curve from the short testing times to the longer commercial cutting times. The agreement between calculated and actual time in this particular instance is much better than would normally be expected, since the plotted values tend not to be strictly colinear, owing to minor variations in the material and the tool. Just as in specifying hardenability by a band rather than a line, so also should the machinability of a material be specified by a band. However, this is not the current practice.

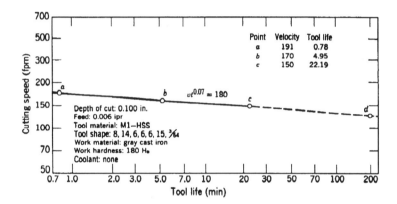

Figure 10-6. Validity of extrapolation of tool life curves. Points a, b, and c were obtained experimentally. A straight line was drawn between them and extrapolated, as indicated by the dashed line. From this line, it was predicted that to obtain a tool life of 200 min, the cutting speed should be 130 fpm. A fourth tool was tested at this speed, with a resulting life of 195 min. This point, d, lies on the extrapolated line.

The equation $vt^n = C$ is valid only when all of the cutting conditions except speed and tool life are kept constant. This restriction includes the tool material and shape, work material, and size of cut. The lack of published data that satisfy the equation for every machining condition often requires the design engineer to modify some existing related data. For example, $vt^{0.1} = 90$ may be the only published data available for machining a certain AISI steel. This relation may have been obtained with HSS tools ground to proper shape with a 1/8 in. depth of cut and 0.012 ipr feed. If carbide cutting tools are being contemplated for this operation, what can one do to obtain some knowledge of the cutting speed–tool-life relationship in this case without actually running a tool-life test? By referring to Table 10-1, one can see that the average value of n for a sintered carbide tool is 0.2 compared to the representative value of 0.1 for a high-speed steel tool. Also, the constant C for the carbide tool is about three to five times as large as the constant for a high speed steel tool. Thus, an approximate expression for the above problem would be $vt^{0.2} = 360$.

In considering the accuracy of such approximations, it is necessary to appreciate the validity of any experimentally determined $vt^n = C$ relationship. Unfortunately, no quantitative studies have been made to indicate the absolute validity of a specific tool-life equation. For ferrous materials, the variation in the machinability of a given type of steel is greater than the variation in the hardness between the upper and lower limits of its hardenability band.

Table 10-1. Representative value for n and C in $vt^n = C$

Tool Material	Range of n	Typical	C
High-speed steel	0.06 - 0.12	0.07	X
Cast nonferrous	0.1 - 0.15	0.12	1.25 to 1.5X
Sintered carbide	0.15 - 0.25	0.2	3 to 5X
Ceramic	0.30 - 0.60	0.4	5 to 8X

In many instances the size of cut has an important effect on the speed-life relationship, and consequently the cost. The cutting speed for a constant tool life, when only the depth of cut and feed are varied independently of each other, is related inversely to the feed and depth exponentially, as follows:

$$v = k_1 f^{-a} \quad \text{and} \quad v = k_2 d^{-b} \qquad (10\text{-}1)$$

By combining these two relations with the speed-life equation, the following useful expression is obtained:

$$v = kt^{-n} f^{-a} d^{-b} \qquad (10\text{-}2)$$

The numerical values of a, as determined experimentally, seem to lie between 0.5 and 0.8 (0.65 typical) and the values of b are in the range of 0.3 to 0.6 (0.40 typical) for both high-speed steel and carbide tools. Representative values of n are given in Table 10-1. Obviously, use of the above relationships requires that n does not vary with the size of cut, when all other operating conditions are held constant. That is, for a given material and tool shape, n has the same numerical value for all values of f and d. This relationship is assumed to be true in the traditional application of the above speed-life equations. However, the more recent studies, discussed in the section Theoretical Machinability, indicate that n *does* vary with the size of cut.

In most industrial operations, the depth of cut is dictated by the initial and final size of the part and is not readily varied during manufacturing. On the other hand, the feed may be varied from small to large values depending on the surface finish requirements and constraints of the setup. From the point of view of costs, because the exponent of f is negative and less than one, it is desirable to machine with as large a feed as is possible since it is apparent from Eq. (10-2) that for any specific increase in feed the cutting speed for a constant tool life is reduced by the ratio of the feed to the a power. That is,

$$v_1 f_1^a = k = v_2 f_2^a \quad \text{or} \quad v_2 = v_1 (f_1 / f_2)^a \qquad (10\text{-}3)$$

Similarly

$$v_2 = v_1 (d_1 / d_2)^b \qquad (10\text{-}4)$$

Since the numerical value of the exponent b is less than the value of a, it is actually more beneficial to use a large depth of cut rather than a large feed, if a choice is available.

It is worthwhile at this time to examine quantitatively the effect upon the rate of metal removal of changing only the feed and the depth of cut. This comparison is made on the basis of maintaining a constant tool life and assuming that n is constant. If the initial conditions are v_1, f_1, and d_1, then the approximate rate of metal removal is

$$Q_1 = 12 v_1 f_1 d_1 \quad \text{in.}^3 / \text{min} \qquad (10\text{-}5)$$

This relationship is more accurate when the diameter of the workpiece is large compared to the depth of the cut.

Upon doubling the feed $(f_2 = 2f_1)$ and without changing the original depth of cut, the new cutting speed for constant tool life (for a = 0.65) is

$$v_2 = v_1 (f_1 / f_2)^a = v_1 (f_1 / 2f_1)^{0.65} = 0.638 v_1$$

The rate of metal removal with the increased feed is

316   Chapter 10

$$Q_2 = 12 v_2 f_2 d_1 = 15.3 v_1 f_1 d_1 \quad \text{in.}^3/\text{min}$$

Thus it is apparent that by doubling the feed and correspondingly reducing the cutting speed to maintain a constant tool life, the rate of metal removal is increased by $(15.3 - 12) \times 100 / 12$, or 27.5 %.

If the depth of cut, rather than the feed, were doubled $(d_2 = 2d_1)$, the new cutting speed would be (for b = 0.40)

$$v_3 = v_1 (d_1 / d_2)^b = v_1 (d_1 / 2d_1)^{0.40} = 0.757 v_1$$

The rate of metal removal with the increased depth of cut is

$$Q_3 = 12 v_3 f_1 d_2 = 18.2 v_1 f_1 d_1 \quad \text{in.}^3/\text{min}$$

The new rate of metal removal is 51.5°% greater than the original rate and 19% greater than when the feed alone is doubled.

In using the above relationships to solve machining problems, when specific values for the constants k and exponents a and b are given, the angles at the tool point must not be varied from those used in obtaining the experimental data. The tool-point shape has a very marked effect on the performance or cutting efficiency of the tool. This effect is discussed in the following section.

TOOL SHAPE AND CUTTING EFFICIENCY

When discussing single-point cutting tools, it is best to describe them by means of six angles and a nose radius, as in Figure 10-7. These have been adopted by the ASA[1], which specifies that they be listed in the following order: back-rake angle, side-rake angle, end-relief angle, side-relief angle, end-cutting-edge angle, side-cutting-edge angle, and nose radius. Each of these angles serves a definite purpose, and a knowledge of their effect upon the cutting speed for a given tool-life is essential for one to become proficient in the engineering applications of the machining processes. Each of these elements will be discussed separately, in order of increasing importance.

---

[1] The American Standards Association, *Single-Point Tools and Tool Posts*, Bulletin No. 522, 1950, p. 3.

Back-Rake Angle

The ordinary range of back-rake angles used on most cutting tools does not affect the cutting speed for a constant tool life. The back-rake is used on a tool primarily to control the direction of chip flow. With a zero back-rake angle, the chip would slide off parallel to the axis of the work in a turning operation; with no chip breaker, the metal would come off in long, straight chips, which are extremely difficult to handle and salvage. A proper back-rake angle will cause the chip to flow at such an angle that it strikes the tool holder or the work material and curls or breaks into small segments.

In cutting steel and other ductile materials, a back-rake angle of approximately 8° works satisfactorily and will cause the chip to curl when cutting at all except the very highest speeds without a special chip breaker ground onto the top of the tool. For more brittle materials, such as cast iron and bronze, a 0° angle is generally used since the material fractures into small irregular chips that break off the work piece in front of the tool face regardless of which angle is used. The most desirable back-rake angle for any cutting operation is the smallest one that will give proper control of the chip, because the time to grind the tool is directly proportional to the magnitude of the angles.

Relief Angles

As their name implies, both the end-relief and the side-relief angles are needed on a tool simply to keep the end and side flanks of the tool from rubbing against the work, which would wear away the cutting-edge. The relief angles are also necessary to enable the tool to be forced into the work material to accomplish the feeding motion. An angle of 6° is usually sufficient, except where a large relief angle would be necessary on the leading side of the tool, as is encountered in thread cutting and similar operations. For hard materials, a slightly smaller relief angle may be used, provided the feed is not large. When cutting with the crater-resistant type of sintered carbide tool, relief angles as large as 12° may be used with better performance than the traditional angle of 6°. Also, for very soft materials and materials with a low modulus of elasticity, slightly better results are obtained by increasing the relief angle up to 12°.

End-Cutting-Edge Angle

The end-cutting-edge angle is provided so that the end of the tool will not contact the work. If it did, the length of the cutting edge would increase to the point of inducing chatter. In a few operations, such as planing cast iron, it is desirable to use a tool with a 0° end-cutting-edge angle to produce flat surfaces

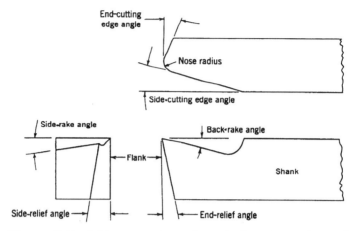

Figure 10-7. ASA nomenclature for a single-point cutting tool. The angles are specified in the following sequence: back-rake, side-rake, end-relief, side-relief, end-cutting-edge, side-cutting edge, nose radius.

free of feed marks. For this type of work, the depth of cut must be very small, on the order of a few thousandths of an inch. However, for the majority of cutting operations, the end-cutting-edge angle should be kept as small as possible, the usual value being 6°. As is true of the relief angles, end-cutting-edge angles of this magnitude have no influence on the relationship between cutting speed and constant tool life.

Side-Rake Angle

For any machining operation, there is an optimal side-rake angle, which is determined by both the work and tool materials. Figure 10-8a illustrates the relationship between the side-rake angle and the cutting speed when machining annealed, medium-carbon, low-alloy steel. By taking the cutting speed for a fixed tool life obtained with a 0° rake angle and, for a basis of comparison, arbitrarily setting it equal to a relative cutting speed of 100%, the relative cutting speed increases to a maximum value of 130%, with a side-rake angle of 25°. As the rake angle increases beyond 25°, the relative cutting speed decreases.

For materials that are softer and more ductile than annealed medium-carbon steel, a rake angle slightly larger than 25° gives the maximum cutting

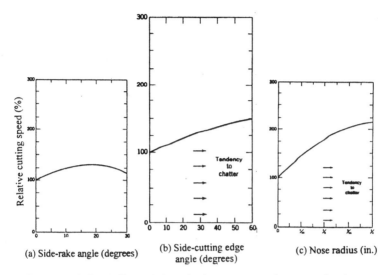

Figure 10-8. Effects of tool shape on cutting speed when machining annealed, medium-carbon, low-alloy steel with HSS tool. Depth of cut = 0.100 in. and feed = 0.12 ipr.

speed, which is also a little greater than 130°. On the other hand, as the hardness of the work material increases, a larger force is placed on the cutting edge. In order to keep this edge from crumbling away, it is necessary to use a somewhat smaller rake angle, with a resulting slight decrease in the maximum cutting speed. In machining alloy steels, especially those strengthened by cold work or heat treating, tools sharpened with a side-rake angle of 5° to 10° perform the most satisfactorily.

Side-Cutting-Edge Angle

Figure 10-8b shows the influence that the side-cutting-edge angle has on the cutting speed for a given tool life when machining a moderately soft steel. An increase in the side-cutting-edge (SCE) angle from 0° to 60° effects an increase of 50% in the relative cutting speed. Because of the tendency of chatter to occur when the length of the cutting edge is extended, it is seldom practical to use angles as large as 60°. However, in some instances when the depth of cut is very small, it is possible to employ a large angle in order to take advantage of the considerable increase in cutting speed that is possible. On most of the roughing work done in job shops, it is possible to use SCE angles of from 20° to 30°.

Nose Radius

The nose radius has the greatest effect of all the elements of the tool point on the maximum cutting speed for any machining operation. Figure 10-8c shows that when turning a moderately soft steel, the cutting speed for a given tool life increases exponentially. The speed for a constant tool life is increased by 120% as the nose radius is enlarged from 0 to 1/4 in. Although a 1/4 in. radius cannot be used extensively on job-shop work, a 1/8 in. radius can be used on much of the roughing and finishing work, with cutting speeds 85% faster than when a sharp-pointed tool is used. In addition to its beneficial effect upon the velocity for a given tool life, the larger nose radius improves the surface finish of the machined work material. Whenever a tool having a small radius is used in conjunction with a large feed, the resulting machined surface resembles a shallow thread.

MACHINABILITY

Machining is primarily a severe plastic deformation process in which the material being cut is locally deformed to fracture. This deformation is accompanied by severe rubbing with high friction on both the tool face at the tool-chip interface and the tool flank at the tool-work contact surface. These three regions are shown in Figure 10-9.

Deforming the material during cutting as well as the friction caused by rubbing produce heat, which makes the tool hot. As it becomes hot, its strength and wear resistance decrease, causing it to "fail," that is, its cutting edge to wear down. Most of the heat generated during machining is carried away with the chips, particularly at high speed, and a small amount is conducted to the work piece. The amount of heat transferred by conduction through the tool to the tool holder and surroundings is negligible. On a thermal basis, machining is essentially a steady-state process within a few seconds after cutting begins.

The term *machinability*, as used by engineers and technicians, in general terms expresses the relative ease of machining a material. The basis for the evaluation or comparison may be any of the following: (1) tool life or cutting speed for a standard cutting condition, (2) cutting forces, (3) horsepower required, (4) surface finish, (5) chip disposal, and (6) dimensional stability. This book uses the first concepts, namely, either the cutting speed for a given tool life or the tool life for a given cutting speed. The concept was first

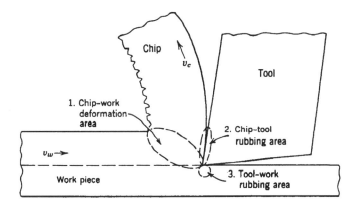

Figure 10-9. The three heat-producing zones in machining.

proposed by Herbert, Rosenhain, and Sturney in the 1920s. It is still the most commonly used concept of machinability since all relative machinability ratings listed in handbooks are based on the cutting speed for a given tool life under some standard cutting conditions (see Table 10-2).

In this context it is apparent that the machining efficiency of a material is dependent not only on the physical properties of the material, but also on the tool material, tool shape, and cutting process. It is extremely useful to have a concept of machinability where it is a characteristic or a secondary property dependent upon primary properties of a material. By this means it is possible to select or evaluate materials during product design on the basis of economic fabrication as well as functional performance.

To achieve a machinability index that is a characteristic of the material, it is necessary to control the variables associated with the tool and cutting conditions. When a material's mechanical properties are evaluated by means of a tensile test, the testing temperature, pressure, environment, and specimen shape are all specified by testing standards. Similarly, when a material's machinability is being evaluated, the testing should comply with specified standards and all of the variables should be stated. When this approach is taken, it is possible to correlate machinability with the material's primary physical properties, as is done in the next section.

To avoid vagueness, it is best to use machinability only when referring to the *characteristic of a material that indicates the cutting speed under standard conditions at which the material can be machined.* Thus machinability will have the units of length over time (ft/min). Whenever other machining parameters, such as cutting forces or surface finish, are being discussed, the term *machinability* should not be used. Instead, the specific parameter under study should be used. Thus, it may be stated that the cutting force during the machining of one material is twice that of a second material; but it would be misleading to imply that the machinability of the second material is twice that of the first.

Table 10-2. Machinability Ratings for Several Materials (AISA B1112 Steel = 100%)

Material	Brinell hardness	Machinability rating (%)
Titanium	250	25
Tool steel	200	30
18-8 Austenitie S.S.	155	30
Monel	200	40
AISI 4340	220	45
AISI 1080	220	45
AISI 1010	130	50
Aluminum bronze	150	60
AISI 1020	140	65
Beryllium	-	65
Malleable iron	140	120
AISI B1113	190	135
Aluminum	-	300 to 1500
Magnesium	-	500 to 2000

The physical properties of a material that affect its machinability fall into two classes: mechanical and thermal. The mechanical properties include the tensile strength, the strain-hardening exponent (or the Brinell hardness, which is a measure of both the tensile strength and the strain-hardening exponent), and the final area ratio or percent reduction of area. All these mechanical properties can be correlated to and predicted from the microstructures of the material, as discussed in Chapters 4 – 7.

By referring to Figure 10-9, the manner in which these physical properties influence a material's machinability may be more easily understood. Heat is generated in the chip-work zone because of the tremendous amount of plastic deformation in that area, and in the tool-chip and tool-work zones because of friction. Since the life of the cutting tool depends upon its cutting-edge temperature, as discussed before, the amount of heat generated and the rate at which it is transferred to the work piece beneath the tool point are very important.

The area under the stress-strain curve, as illustrated in Figures 7-4 and 7-6, is an indication of the work (per unit volume) done on a material when it is plastically deformed to fracture, as in machining. This area may be determined if the tensile strength, the strain-hardening exponent, and the percent reduction of area are known. Obviously, the larger this area, the greater the amount of heat generated, with resulting higher tool-cutting edge temperatures. Thus, in terms of its mechanical properties, a material will have a high machinability when these properties are such that the stress-strain area is small.

The thermal properties of a material that influence its machinability are: density ρ, lb/ft^3; heat capacity C_p, Btu/lb°F; and thermal conductivity k, Btu/ft/min/°F. The product of the density and the specific heat at constant pressure is equal to the volume specific heat, that is, $C_v = \rho C_p$. These three properties may be combined into one, the thermal diffusivity α, which is $k/\rho C_p$ and has the units of ft^2/min..

In the machining process, the heat capacity of the work material determines the amount of heat a given mass can absorb beneath and in front of the chip. The thermal conductivity of the work material determines the rate at which this heat is added to the given mass. The thermal diffusivity, on the other hand, is an inverse measure of the time required to heat this mass to a specified temperature. Thus when machining a material having a high thermal diffusivity, the work material in front and beneath the chip will become hot in a relatively short time, with a resulting lower cutting edge temperature.

In addition to influencing the machinability of a material, the mechanical properties also help determine the type of chip formed during machining (Figure 10-10). When machining a brittle material, that is, one having an area reduction of nearly 0%, there is practically no plastic deformation in the chip-work zone, and consequently, as the tool progresses, it causes the material in its path to fracture into small, irregularly shaped, undeformed chips as shown in Figure 10-10a. Chips of this type are desirable from a material handling, storing, and transporting point of view in that the chips have a relatively high bulk density when compared to the stringy type of chips that are formed from ductile materials.

How thick and thin chips are formed when machining a ductile material is shown in Figure 10-9b and c. Both of these chips are referred to as the continuous type because they come off in long strings or coils. In b the shear angle ϕ is low, and consequently the thickness of the chip is greater than the thickness of the cut h. In this case ($\phi = 20°$) the chip-thickness ratio, t/h is 3.

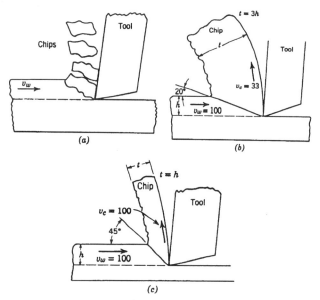

Figure 10-10. Types of chips formed in machining. (a) Discontinuous chips formed from brittle material. (b) and (c) Continuous chips formed with two materials of different ductility. (b) $\theta = 20°$. (c) $\theta = 45°$ (the ratio of t/h is called the chip-thickness ratio).

In other words, the length of the chip is one-third the length of the cut inasmuch as the volume of the material remains constant. The chip shown in c has a thickness equal to the thickness of the cut, or a cutting ratio of one. Unfortunately, the chip-thickness ratio, or the cutting ratio as it is sometimes called, cannot be determined from the mechanical properties, but must be determined by an actual machining test.

A FUNDAMENTAL MACHINABILITY EQUATION

A comprehensive study of the influence of the physical properties of a material on its machinability has been made by Henkin and Datsko.[2] This investigation yielded a relationship that equates the machinability of a material to the physical properties of thermal conductivity, Brinell hardness, and percent reduction of area. The derivation of this fundamental machinability equation is presented below.

After considerable experimental and analytical study it was determined that seven independent variables, in addition to the tool-chip interface temperature, are required to define machinability. These eight variables are listed in Table 10-3. Many other parameters characterize the machining process, such as cutting forces, coefficient of friction, "shear" angle, chip ratio, and abrasion. However, all these additional parameters are really dependent upon the primary variables listed in Table 10-3, although the exact nature of this dependency is not known.

Investigation of all the mechanical properties, including friction and wear resistance, has shown that the Brinell hardness and the area ratio at fracture (or percent reduction of area) describe sufficiently well the plastic behavior of a material for the prediction of machinability. As far back as the 1920s, E. G. Herbert came to the conclusion that "the measure of machinability is the hardness of the chip."

Table 10-3. Variables Used in Dimensional Analysis of Machinability

Variable	Dimensions	Descriptions
H_B	$ML^{-1}t^{-2}$	Brinell hardness (kg/mm^2)
R_f	None	Area ratio at fracture
k	$MLt^{-3}T^{-1}$	Thermal conductivity (Btu/hr-ft-°F)
C_v or ρC_p	$ML^{-1}t^{-2}T^{-1}$	Volume specific heat (Btu/ft^3-°F)
h	L	Thickness (feed) of cut (in.)
w	L	Width (depth) of cut (in.)
v	Lt^{-1}	Cutting velocity (ft/min)
T	T	Tool-chip temperature (°F)

Hardness implies resistance to penetration or compressive deformation. The range of elastic deformation for materials, except for elastomers and the like, is quite small and therefore hardness is primarily influenced by the plastic

---

[2]Henkin, A., and J. Datsko, The Influence of Physical Properties on Machinability, *Trans. ASME 85*, Series B, No. 4.

properties of the material: namely, the stress coefficient σ_o and the strain-strengthening index m in the equation $\sigma = \sigma_o \varepsilon^m$. Inasmuch as the average or effective strain a material is subjected to beneath a Brinell penetrator during a hardness test is about the same as during a tensile test, the Brinell hardness provides sufficient information of the plastic portion of a material's behavior. It is well known that the tensile strength is proportional to the Brinell hardness and that the proportionality constant is 500 for steel. For the analysis that follows, it must be recalled that the Brinell hardness has the units of stress, kg/mm^2, or energy per unit volume.

In order to determine the area under the stress-strain curve, which is a measure of the work per unit volume necessary to deform a material to fracture, the tensile strain at fracture, ε_f, must also be known. The strain at fracture may be determined from either the final area ratio or the percent reduction of area by means of the following relationship which was developed in Chapter 7.

$$\varepsilon_f = \ln R_f = \ln \frac{100}{100 - A_r} \qquad (10\text{-}6)$$

The thermal properties that influence machinability are the volume-specific heat ρC_p and the thermal diffusivity α. These have a significant effect, provided that the heat transfer through the tool as well as the heat convected to the surroundings are of second-order importance. Experimental data verify the role of these properties in dry cutting, that is, machining without a cutting fluid.

The tool life also depends upon the cutting velocity v, the thickness of cut or feed f, and the depth of cut d. As mentioned previously, parameters such as tool shape and tool material, which influence the tool life, are considered to be constants in this analysis, so the resulting relationship is a reflection only of the material's properties.

The cutting edge temperature T is a dependent variable, the magnitude of which is determined by the cutting velocity. As can be seen from the experimental data plotted in Figures 10-11 and 10-12 for a wide range of materials tested, the cutting edge temperature is 975 ± 25°F for a cutting velocity that results in a 60 minute tool life, designated as v_{60}. Similarly, for a v_{20} the cutting edge temperature is found to be 1025 ± 25°F in Figures 10-11 and 10-12. Note that it is not necessary to include time in the fundamental machinability relationship.

The Pi-theorem of dimensional analysis states that the number of dimensionless groups needed to completely describe a physical phenomenon is equal to the difference between the total number of variables and the number of

Machinability of Metals 327

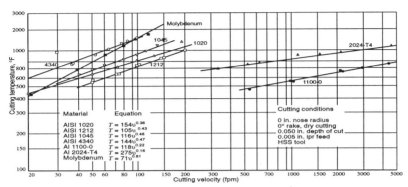

Figure 10-11. Relationship between cutting speed and tool-chip interface temperature.

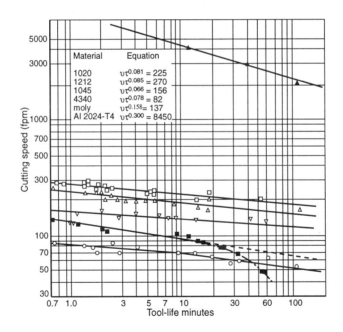

Figure 10-12. Relationship between cutting speed and tool life for some metals.

primary variables. Four primary variables are used in this analysis: length L, time t, mass M, and temperature T. As it has been determined by experimental and analytical means that the eight variables listed in Table 10-3 are the total number needed to describe machinability, the number of dimensionless groups should be four.

Dimensionless analysis by itself does not indicate what the four dimensionless groups are. After an extensive study, it was established that the four required groups were: vf/α, d/f, $\rho C_p TJ/H_B$, and R_f. (α is the thermal diffusivity, $k/\rho C_p$; J is the mechanical equivalent of heat; R_f is the area ratio at fracture, A_o/A_f.)

The first group (vf/α) is the so-called thermal number, whose importance was first recognized by Bisacre and Chao[3]. The thermal number governs the ratio of the heat conducted through the work to the heat transported by the moving chip. When machining with a large v and a small α, most of the heat is carried away with the chips. But with a small v and large α, most of the heat goes into the work piece.

The second group (d/f) represents the geometry of the cut and is discussed in detail in the following paragraphs, where the exponents for these dimensionless groups are derived.

The third dimensionless group $\rho C_p TJ/H_B$ can be considered to represent a temperature rise if all the heat is carried away with the chip and none is conducted to the work piece.

The fourth group, the area ratio R_f or A_o/A_f, is a measure of the work done on a material during machining, and consequently the machinability is inversely related to this group. At the limit when $A_r = 100\%$ and $R_f = \infty$ or, in other words, when a material can withstand an infinitely large strain, the material should be unmachinable.

These four dimensionless groups may be equated as follows:

$$\left(\frac{v_x f}{\alpha}\right)\left(\frac{d}{f}\right)^b = K_p \left(\frac{\rho C_p T_x J}{H_B}\right)^c \left(R_f\right)^d \qquad (10\text{-}7)$$

where the subscript x refers to a particular tool life and corresponding cutting edge temperature. K_p is a proportionality constant that is a function of the tool

---

[3] Bisacre, G. H., and B. T. Chao, The Effect of Speed and Feed on the Mechanics of Metal Cutting, *Proceedings Inst. Mech. Engr.* 154, 1951, pp.1-13.

shape and tool material. To determine the exponents in this equation, actual machining tests were run on a range of experimental materials, including tool life and cutting temperature as a function of cutting velocity. These were then correlated with the physical properties obtained for the same materials that were machined.

In order to evaluate the machinability of materials, it is necessary to select one set of standard conditions to serve as a basis, as is true of all physical tests. A convenient basis for establishing machinability, one that has frequently been used, is the cutting speed that results in a 60 min tool life, v_{60}. As mentioned previously, this corresponds to a cutting edge temperature of about 975°F, with the temperature increasing for a shorter tool life.

By carefully examining the deformation and shearing of the material in front of the tool and observing the process of machining metallurgically polished and etched specimens through a microscope, one can see that most of the deformation occurs in the region where the temperature of the material is between room temperature and the cutting edge temperature. By correlating the experimental tool-life data with these microscopic studies, it can be established that most of the deformation occurrs at a temperature slightly above the average of the room temperature and the cutting edge temperature. On that basis, when machining with a v_{60}, the physical properties of the material used in the machinability equation to calculate the machinability must be obtained at a testing temperature of approximately 600°F.

By plotting the 600°F physical properties versus the experimental machinability data, the numerical values of the exponents c and d were found to be 1 and –0.5, respectively.

To determine the effect of the size of the cut on the constant-life velocity, it is necessary to consider the two dimensionless groups vf/α and $(d/f)^b$. For a better understanding of the effect of the size of cut, these two groups can be combined into one, $v/\alpha(f^{1-b}d^b)$, where $(f^{1-b}d^b)$ may be considered as a "characteristic length" and designated as q. It should be noted that the exponent of q is unity and the expression for q can be expressed in the more commonly used form $(f^a d^{1-a})$. From experimental data as well as from physical considerations, it is apparent that a is not a constant having one value, but instead it varies with the depth-to-width ratio of the size of the cut. The value of a can easily be determined from published experimental data for a range of values of the d/f ratio.

It is known from machining experience that the depth of cut, when it is extremely large in comparison to the feed, has practically no effect on the constant-life velocity. Expressed mathematically: as the ratio of d/f goes to infinity, the limit of a will be 1. Also, when f = d, the effect of each on the

330   Chapter 10

constant-life velocity obviously should be the same and therefore a will be 1/2. These two limiting values are in agreement with published experimental data.

The value of a for d/f values between ∞ and 1 can be established by studying the published[4,5] experimental values of the exponents for the feed and depth of cut. The results of many of the experimental studies are summarized with an empirical equation of the form

$$v_x = Cf^{-a}d^{-b} \qquad (10\text{-}8)$$

where a and b are about 2/3 and 1/3, respectively, for average-size turning cuts. Analysis of thousands of experimental values for the exponents a and b, shows that they are a function of the feed over depth ratio. Further study has found that this relationship can be expressed as

$$a = 0.6(f/d)^{-0.05} \qquad (10\text{-}9)$$

By rewriting Eq. (10-7) explicitly for v_x; we get

$$v_x = K_p \left(\frac{\alpha \rho C_p J T_x}{q H_b} \right) (R_f)^{-0.5} \qquad (10\text{-}10)$$

Since the product of $\alpha \rho C_p$ is equal to the thermal conductivity k and R_f is equal to e^{ε_f} we can rewrite Eq. (10-10) as

$$v_x = K_p \left(\frac{kJT_x}{qH_B} \right) e^{-0.5\varepsilon_f} \qquad (10\text{-}11)$$

One of the shortcomings of the constants C or K in Taylor equations is that the work material properties and tool material and shape are all included in the same constant. For engineering analysis and applications, it is desirable to separate the tool from work materials properties. This can be accomplished by grouping the work material properties into one constant, such as

---

[4] ASME, *Manual on Cutting of Metals*, American Society of Mechanical Engineers, New York, 1952.
[5] Boston, O. W. *Metal Processing*, John Wiley & Sons, New York, 1952.

$$B = k \frac{\left(e^{-0.5\varepsilon_f}\right)}{H_B} \tag{10-12}$$

and tool and cutting conditions into a separate constant

$$A = K_p JT_x \tag{10-13}$$

so that Eq. (10-7) can be rewritten as

$$v_x = \frac{A_x B}{q} \tag{10-14}$$

It has been determined that the size of cut constant q is also influenced by the strain-strengthening capacity, m, of the work material. By analyzing thousands of experimental data points, the author found that the effect of the work-hardening ability of the metal can be expressed by the relationship

$$q = (f+y)^a (d+y)^{1-a} \tag{10-15}$$

where the work-hardening factor y can be determined from the expression

$$y = 0.0055 m^{0.5} \tag{10-16}$$

Finally, extensive study of the tool and cutting condition factors has yielded the following relationship, which includes all of the variables that affect the cutting speed-tool life relationship except for the work material and size of cut.

$$A_X = A_T \cdot A_M \cdot A_E \cdot A_C \cdot A_F \cdot A_H \tag{10-17}$$

For the tool life constant where t is the tool life in minutes:

$$A_T = 30(t)^{-0.07} \tag{10-18}$$

For the tool material constant:

$$A_M = 1.0 \text{ for T-1 or M-1 H.S.S. tools} \tag{10-19.a}$$

332   Chapter 10

$$A_M = 4t^{-0.14} \quad \text{for} \quad WC \tag{10-19.b}$$

$$A_M = 6t^{-0.32} \quad \text{for} \quad C \tag{10-19.c}$$

For the effective rake angle constant:

$$A_E = 0.9 \quad \text{for} \quad ER/ERO < -0.2 \tag{10-20.a}$$

$$A_E = 1 + (0.6 - 0.001\,H_B) \times ER/ERO$$
$$\text{for} \quad -0.2 < ER/ERO < 1 \tag{10-20.b}$$

$$A_E = 1.8 - 0.00133\,H_B - (0.2 - 0.00033\,H_B) \times ER/ERO$$
$$\text{for} \quad 1 < ER/ERO < 4 \tag{10-20.c}$$

$$A_E = 1.0 \quad \text{for} \quad ER/ERO < 4 \tag{10-20.d}$$

where ER is the effective rake angle, ERO is the optimal effective rake angle, and H_B is the Brinell hardness at 600°F. $ERO = 46 - 0.1\,H_B$ for HS.

For the length-of-cutting-edge constant:

$$LCE = R(1.571 - 0.017\,SCE) + [d - R(1 - \sin SCE)] / \cos SCE \tag{10-21}$$

$$A_c = (LCE/d)^{0.67} \tag{10-22}$$

where R is nose radius, d is the depth of cut, SCE is the side cutting edge angle, and LCE is the length of the cutting edge.

For the cutting fluid constant:

	Dry	Heavy Oil	Light Oil	Water Base
A_F	1.00	1.10	1.15	1.25

For the workpiece size constant:

$$A_h = (h/2)^{0.25} \tag{10-23}$$

where h is the radius of a solid bar, the wall thickness of tubing or the thickness of a plate.

Machinability of Metals   333

EXAMPLE PROBLEMS

10-1. A lathe operator is machining AISI 1018 hot-rolled steel axles and is reducing the diameter from 3.500 in. to 3.100 in. in one cut with a properly ground tool. He is using a feed of 0.006 ipr at a speed of 60 rpm, and the length of cut is 12 in. The loading-unloading time is 1 min, and the tool changing time is 1 min. Machinability studies show that for this tool-work combination, the following relationship is valid.

$$v = 2t^{-0.1}f^{-0.7}d^{-0.4}$$

After this turning operation, the shafts are ground to 3.090 in. diameter.

Assuming that a tool life of 30 min is satisfactory, by what percentage could the daily productivity be increased by specifying a change in cutting conditions? (420 working min in a day). (There is not one best answer, but several correct answers.)

Solution:
This is a roughing cut and a large feed can be used.
Initial conditions: $v = (\pi \times 3.5 \times 60)/12 = 54.8$ fpm
$t = 2.48^{10} = 8800$ min; thus tool changing time/pc is nil.
Total time/shaft: $t_c = L/(f \times N) = 33.3$ min and $t(load) = 1$ min;
tot.time $= 33.3 + 1 = 34.3$ min.
Production $= 420/34.3 = 12.2$ axles/day.
Solution #1. Same f and d, but $t = 30$ min.
$v = 96.6$ fpm; $N = 105$ rpm; $FR = 0.006 \times 105 = 0.63"/$min
$t_{per\ shaft} = 12"/0.63/min= 19$ min; so change tool after each shaft.
Tot. time $= 19 + 1 + 1 = 21$ min; shafts/day $= 420/21 = 20$.
% increase $= 63.9\%$

Solution #2. Use $f = 0.020$ ipr and $t = 30$ min; $v = 41.7$ fpm;
To prevent tool failure while cutting a shaft, change tool after 2 shafts.
Then tct/pc $= 0.5$ min. Tot. time $= 13.2 + 1 + 0.5 = 14.7$ min.
Shafts/day $= 420/14.7 = 28.5$. % increase is 133%.

Solution #3. Using $f = 0.04$ ipr, $t_c = 10.7$ min, $t_{tc} = 0.356$ min

334 Chapter 10

Tot. time = 10.7 + 0.356 + 1 = 12.06 min; shafts/day = 420/12.06 = 34.8;
% increase = 185%.

10.2 In the absence of any experimental data, estimate the value of v_{60} for turning a 4" diameter bar of a new alloy that has the following properties: k = 12 BTU/hr ft°F, S_y = 220 ksi, S_u = 245 ksi, e_u = 0.01, e_f = 0.65. The properties at 600°F are only slightly different from the room temperature values. A HSS tool is used that has a 0°SCEA, 10°SR, 1/16" radius. The cutting conditions are: 0.100" depth of cut, 0.010 ipr feed and no coolant. Also, how much faster could a proper carbide tool be run?

Solution:
Assume H_B = 492, d = 2.76, K_B = 0.535
  Check S_u: 0.535 × 492 = 263 > 246
Assume H_B = 470, d = 2.82, K_B = 0.525, S_u = 0.525 x 470 = 247
∴ H_B = 470

$B = (12/470)e^{-0.325} = 0.018$

$a = 0.6(0.01/0.1)^{-0.05} = 0.67 \quad y = 0.0055(0.01)0^{.5} = 0.0006$

$q = (0.01+.0006)^{0.67}(0.1+.0006)^{0.33} = 0.022$

$AT = 30(60)^{-0.07} = 22.52$
AM = 1 for HSS
ERO = 46 –47 = –1
ER = 10°   ER/ERO = –10
AE = 0.9
LCE = 1/16 (1.571– 0.017×0) + [0.1– 0.0625 (1– sin 0°)]/cos0°
    = 0.136"
$AC = (0.136/0.1)^{0.67} = 1.227$
AF = 1.00 (Dry)
$AH = (2/2)^{.25} = 1.0$ for 4" dia.
AX = 22.52 × 1.0 × 0.9 × 1.227 = 24.9
v_{60} = 24.9 × 0.018 / 0.022 = 20.4 fpm

for carbide

$$AM = 4(60)^{-0.14} = 2.25 \quad v_{60} = 2.25 \times 20.4 = 45.8 \text{ fpm}$$

Note: An alternate way to convert from HSS tools to WC tools is to use the Taylor equation $vt^n = C$.

For HSS: $20.4 \times 60^{0.07} = 27.2 = C_{HSS}$

$C_{wc} = 4 \times C_{HSS} = 4 \times 27.2 = 108 \text{ fpm}$

$n_{wc} = 0.07 + 0.14 = 0.21$

For WC tool: $V_{60} = 108 \times 60^{-0.21} = 45.7 \text{ fpm}$

REFERENCES

1. Boston, O.W. *Metal Processing,* John Wiley & Sons, New York, 1951.
2. ASME, *The Manual on Cutting of Metals,* Am.Soc. of Metals, New York, 1952.
3. Datsko, J., *Material Properties and Manufacturing Processes,* John Wiley & Sons, John Wiley & Sons, New York, 1965.
4. Henkin, A. and J. Datsko, *The Influence of Physical Properties on Machinability,* Trans. ASME, Series B, Vol. 4, 1963.

Chapter 10

STUDY PROBLEMS

10-1. A process sheet specifies the following tool shape for rough turning AISI 5140 annealed shafts that are 2 in. diameter and 12 in. long: 5° BR, 0° SR, 6° ECA, 6° SCA, 6° ECEA, 0° SCEA, 0 in. R. By what percent could the cutting time be reduced simply by changing the tool shape? Explain.

10-2. You are asked to specify the optimal cutting speed and tool life for a specific machining operation in which sintered carbide tools will be used. In looking through the handbooks, the only machinability data you can find for this particular material (hot-rolled 9140) is:

$$v = 1.8t^{-0.1}f^{-0.8}d^{-0.4}$$

which pertains to machining with a HSS tool having the proper tool shape. Rewrite the above equation so that it will express the cutting speed-tool life relationship when cutting with a sintered carbide tool with 1/4 in. depth of cut and 0.020 ipr feed.

10-3. Compare for two cases (a and b) the machinability (cutting speed tool life) of AISI 1020 HR to AISI 1020 CD (10% CW) steel. Compare these values to published data.
For 1020 HR, use $H_B = 116$, $Su = 58$ksi, $m = 0.25$, $\varepsilon_f = 0.913$, $k = 25.5$ BTU / hr ft °F.

10-4. Compare the machinability ratings (B/q ratio) of annealed 304 stainless steel to annealed 17-4 PH stainless steel. The properties are:

	k BTU/hr.ft.°F	S_y ksi	S_u ksi	m	ε_f	H_B kg/mm²
304	9.4	34	93.3	0.57	1.43	156
17-4 PH	11.5	122	153	0.05	0.86	310

a. d = 0.100" and f = 0.002 ipr
b. d = 0.100" and f = 0.020 ipr

Machinability of Metals 337

10-5 Compare the metal removal rate, Q(in.^{3}/min) and length of cut (in.) in one minute when turning a 4" dia. bar of 304 stainless steel. A carbide tool having an A_{60} value of 90 is being used. Consider two cases:

1. depth of cut of 0.100" and feed of 0.002 ipr
2. depth of cut of 0.100" and feed of 0.02 ipr

Note: From problem 10-4 the B/q ratios for 304 stainless steel are 2.85 for case 1 and 0.93 for case 2.

10-6. One Air Force sponsored machinability project conducted by Boeing Company reports that the nickel-base alloy Rene 41 is 50% more machinable than the cobalt-base alloy HS25. A second Air Force sponsored machinability project reports that HS25 alloy is 50% more machinable than Rene 41. Both are for HSS tools with a depth of cut of 0.100" and feed of 0.006 ipr. The second report also gives the following properties for each (600°F):

Matl.	S_u ksi	S_y ksi	m	Ar	Rc	k BTU/hr.ft°F
HS25	140	50	0.50	34	20	9.0
Rene 41	178	110	0.32	44	35	8.5

Which is probably more machinable?

Note: You must first evaluate the listed properties as best you can.

10-7. A U.S. Air Force Machinability Report recommends three equations to calculate the cutting speed for a 60 minute tool life with HSS tools, and the strain strengthening exponent m, on the basis of the work material's physical properties.

(32) $m = (0.396\, Su/Sy)^{3.25}$

(98) $m = 1.117 + 0.00414\, Su - 0.00736\, Sy - 0.39634\, Sy/SU$
$- 0.03324\, E + 0.011\, E1 + 0.0029\, Ar + 0.03341\, k$

(101) $v_{60} = 216.27 + 139.80315\, m - 42.80513\, Sy/Su - 7.32979\, E$
$- 2.09984\, E1 + 2.02145\, Ar$

(102) $v_{60} = -31.05 + 446.64673\, m + 2.15589\, Sy - 9.72155\, E$
$+ 1.39505\, Ar - 8.34685\, k$

Chapter 10

(103) $\quad v_{60} = 292.73 - 1.23230 \, Sy - 5.76956 \, E - 2.81411 \, El$
$\qquad + 1.88939 \, Ar + 5.14136 \, k$

The report states that all equations predict almost perfectly the v_{60}'s and m value for any metal. On the basis of engineering principles and a knowledge of the machining mechanism, critically evaluate each of the equations.

Appendix A

Conversion Factors: U.S. Common to S.I. Units

Multiply	by	To Obtain*	Multiply	by	To Obtain*
(Energy)			**(Pressure or Stress)**		
Btu	1.054×10^3	J	kg(force)/mm^2	9.807×10^6	N/m^2
Calorie	4.187	J	ksi	6.895×10^6	N/m^2
ft lbf	1.356	J	Pa	1.000	N/m^2
kwhr	3.600×10^6	J	**(Velocity)**		
W-s	1.000	J	fpm	5.080×10^{-3}	m/s
(Force)			**(Thermal Conductivity)**		
kg	9.807	N	Btu/hr ft°F	5.70×10^{-2}	W/mK
lb	4.448	N	cal/scm°C	4.187×10^4	W/mK
(Length)			**(Thermal Diffusivity)**		
Å	1.000×10^{-10}	m	ft^2/hr	2.581×10^{-5}	m^2/s
in.	2.540×10^{-2}	m	**(Specific Heat Capacity)**		
in.^{2}	6.452×10^{-4}	m^2	Btu/lb°F	4.187×10^3	J/kg K
in.^{3}	1.639×10^{-5}	m^3	cal/g°C	4.187×10^3	J/kg K
(Mass)			Btu/lb	2.324×10^3	J/kg
lb	4.536×10^{-1}	kg	**(Temperature)**		
(Power)			°F + 459.67	5.556×10^{-1}	K
Btu/hr	2.931	W	°C + 273.15	1.000	K
ft lbf/s	1.356	W			
hp	7.457	W			
(Density)					
lb(mass)/in.^{3}	2.768×10^4	kg/in.^{3}			

* J=joule, N=newton, m=meter, kg=kilogram, W=watt, Pa=pascal, s=second, A=ampere, K=kelvin, C=Celsius m^2

Appendix B

Table B-1. Physical Properties of Metals

Material	Lattice	Density ρ, g/cm^3	Specific heat C_p, cal/sec g°C	Thermal conductivity k, cal/sec cm°C	Thermal diffusivity α, cm^2/sec	Melting temperature, °F	Modulus of elasticity E, psi ×10^6	Yield strength S_y, psi	Tensile strength S_u, psi	Elongation %	Reduction of area %	Heat of fusion, ΔH_f cal/g
Steel: Ferrite	BCC	7.86	0.11	0.15	0.173	2780	30	30,000	40,000	45	75	58
Steel: 1020	—	7.86	0.11	0.15	0.173	2770	30	40,000	60,000	40	60	58
Steel: 4340	—	7.84	0.120	0.08	0.085	2750	30	135,000	153,000	10	35	58
Al: 99.996	FCC	2.70	0.22	0.53	0.89	1218	10	1,800	7,000	50	—	77
Al: 2S-0	FCC	2.71	0.23	0.53	0.88	1215	10	5,000	13,000	45	—	77
Zirconium	HCP	6.5	0.066	0.07	0.16	3200	12	16,000	36,000	30	—	f60
Tantulum	BCC	16.6	0.036	0.13	0.22	5425	27	23,000	50,000	40	—	38
Columbium	BCC	8.57	0.065	0.13	0.23	4380	15	30,000	40,000	30	80	69
Beryllium	HCP	1.82	0.516	0.385	0.41	2340	40	26,500	33,000	1	—	260
Hafnium	HCP	13.09	0.0351	0.0533	0.116	3100	20	33,600	64,800	23	37	—
Molybdenum	BCC	10.2	0.061	0.35	0.56	4760	50	60,000	150,000	10	50	70
Copper	FCC	8.96	0.092	0.94	1.14	1981	16	8,000	32,000	55	78	42
70-30 Brass	FCC	8.5	0.09	0.25	0.33	1740	16	16,000	48,000	65	75	—
304 S.S.	FCC	7.9	0.12	0.039	0.054	2600	25	33,000	86,000	60	75	—
Titanium	HCP	4.54	0.126	0.0685	0.120	3300	16.8	63,000	79,000	25.2	50	97
Nickel	FCC	8.9	0.105	0.22	0.236	2651	30	8,500	46,000	30	70	74
Ni base alloy[1]	FCC	8.25	0.108	0.575	0.645	2300	31.7	154,000	206,000	14	—	—
Co base alloy[2]	FCC	9.13	0.092	0.27	0.322	2500	32.9	70,000	150,000	65	40	—

[1] 19 Cr, 11 Co, 10 Mo, 3 Ti, Bal Ni
[2] 10 Ni, 20 Cr, 15 W, Bal Co

Conversion factors: 1 cal = 3.968×10^{-3} Btu; 1 cal/g = 1.8 Btu/lb; 1 cal/g °C = 1 Btu/lb °F

Appendix B   341

Table B-2. Tensile Properties of Some Metals[3]

Material	Condition	Yield Strength (ksi)	Tensile Strength (ksi)	σ_o (ksi)	m	ε_f
Carbon and Alloy Steels[a]						
1002	1500°F-1 hr A 0.032"	22.0	39.5	76.0	0.29	1.25
1002[4]	1800°F-1 hr A	19.0	42.0	78.0	0.27	1.25
1008 DQ	as rec'd 0.024"	25.0	39.0	70.0	0..24	1.20
1008 DQ	as above–trans	27.0	43.0	70.0	0.24	1.10
1008 DQ	1600°F-1 hr A	26.5	40.0	—	—	—
1010	0.024" CRS strip	33.2	47.5	84.0	0.23	1.20
1010	as above–trans	36.8	48.5	88.0	0.26	1.00
1010	1600°F-1 hr A	28.6	44.2	82.0	0.23	1.20
1010	as above–trans	29.1	43.8	82.0	0.23	1.20
1018	A	32.0	49.5	90.0	0.25	1.05
1020	HR	42.0	66.2	115.0	0.22	0.90
1045	HR	60.0	92.5	140.0	0.14	0.58
1144	A	52.0	93.7	144.0	0.14	0.49
1144[5]	A	50.0	93.7	144.0	0.14	0.05
1212	HR	28.0	61.5	110.0	0.24	0.85
4340	HR	132.0	151.0	210.0	0.09	0.45
52100	spher A	80.0	101.0	165.0	0.18	0.58
52100	1500°F A	131.0	167.0	210.0	0.07	0.40
Stainless Steels						
18-8	1600°F-1 hr A	37.0	89.5	210.0	0.51	1.08
18-8	1800°F-1 hr A	37.5	96.5	230.0	0.53	1.38
302	1800°F-1 hr A	34.0	92.4	210.0	0.48	1.20
303	A	35.0	87.3	205.0	0.51	1.16
304	A	40.0	82.4	185.0	0.45	1.67
202	1900°F-1 hr A	55.0	105.0	195.0	0.30	1.00
17-4 PH	1100°F aged	240.0	246.0	260.0	0.01	0.65
17-4 PH	A	135.0	142.0	173.0	0.05	1.20
17-7 PH	1050°F aged	155.0	185.0	225.0	0.05	0.90
17-7 PH	900°F aged	245.0	255.0	300.0	0.04	0.50
440 C	Solution HT	63.5	107.0	153.0	0.11	0.36
440 C	A 1600°F-50°F/hr	67.6	117.0	180.0	0.14	0.12

[3] All values are for longitudinal specimens except as noted. These are values obtained from only 1 or 2 different heats. The values will vary from heat to heat because of differences in composition. ε_f may vary by 100%.
[4] 3/4" dia. bar.

Table B-2 (cont'd.)

Material	Condition	Strength Yield (ksi)	Tensile (ksi)	σ_o (ksi)	m	ε_f
		Aluminum Alloys				
1100	900° F-1 hr A	4.5	12.1	22.0	0.25	2.30
3003	800° F-1 hr A	6.0	15.0	29.0	0.30	1.50
2024[5]	T-351	52.0	68.8	115.0	0.20	0.37
2024	T-4	43.0	64.8	100.0	0.15	0.18
7075	800° F A	24.3	33.9	61.0	0.22	0.53
7075	T-6	78.6	86.0	128.0	0.13	0.18
2011	800° F-1 hr A	7.0	25.2	41.0	0.18	0.35
2011	T-6	24.5	47.0	90.0	0.28	0.10
		Magnesium Alloys				
HK31XA	800° F-1 hr A	19.0	25.5	49.5	0.22	0.33
HK31XA	H24	31.0	36.2	48.0	0.08	0.20
		Copper Alloys				
ETP Copper	1000° F-1 hr A	4.7	31.0	78.0	0.55	1.19
ETP Copper	1250° F-1 hr A	4.6	30.6	72.0	0.50	1.21
ETP Copper	1500° F-1 hr A	4.2	30.0	68.0	0.48	1.26
OFHC Copper	1250° F-1 hr A	5.3	33.1	67.0	0.35	1.00
90-10 Brass	as rec'd[5]	12.8	38.0	85.0	0.43	—
90-10 Brass	1200° F-1 hr A	8.4	36.4	83.0	0.46	—
90-10 Brass	as above + 10% CW + 1200° F A	6.9	35.0	87.0	0.51	1.83
80-20 Brass	1200° F-1 hr A	7.2	35.8	84.0	0.48	—
80-20 Brass	as above + 10% CW + 1200° F A	6.4	34.6	85.0	0.52	1.83
70-30 Brass	1200° F-1 hr A	12.1	44.8	112.0	0.59	—
70-30 Brass	as above + 10% CW + 1200° F A	10.7	43.4	107.0	0.59	1.62
70-30 Brass[4]	1000° F-1 hr A	11.5	45.4	110.0	0.56	1.50
70-30 Brass[4]	1200° F-1 hr A	10.5	44.0	105.0	0.52	1.55
70-30 Brass[4]	1400° F-1 hr A	8.8	42.3	105.0	0.60	1.60
70-30 Leaded Brass	1250° F-1 hr A	11.0	45.0	105.0	0.50	1.10
Naval Brass[6]	1350° F-1/2 hr A	17.0	54.5	125.0	0.58	1.00
Naval Brass[6]	1350° F-1/2 hr Wq	27.0	66.2	135.0	0.37	0.50
Naval Brass[6]	850° F-1/2 hr A	17.5	56.0	125.0	0.48	0.90
Naval Brass[6]	850° F-1/2 hr WQ	31.5	64.5	135.0	0.37	0.80
Naval Brass[6]	1500° F-3 hr A	11.0	48.0	—	—	0.74

[5]Tensile specimen machined from 4" dia. bar transverse to rolling direction.
[6]Specimens cut from 1/2" hot-rolled plate.

Appendix B 343

Table B-2 (cont'd.)

Material	Condition	Yield Strength (ksi)	Tensile Strength (ksi)	σ_o (ksi)	m	ε_f
		Nickel Alloys[7]				
Nickel 200	1700° F-15 min WQ	16.2	72.	150.0	0.375	1.805
Nickel 99.44%[7]	CD + A	20.5	73.7	160.0	0.40	1.47
Monel 400	1700° F-15 min WQ	26.5	77.7	157.0	0.337	1.184
Monel K500	1700° F-15 min WQ	34.4	92.6	182.0	0.32	1.305
Inconel 600	1700° F-15 min WQ	46.6	102.5	201.0	0.3315	1.14
Inconel 625	1700° F-30 min WQ	77.1	139.7	297.0	0.395	0.75
Inconel 718	1750° F-20 min WQ	43.6	99.4	205.0	0.363	1.337
Inconel 750	2050° F-45 min WQ	36.4	106.5	230.0	0.415	1.27
Incoloy 800	2050° F-2 hr AC	22.2	77.1	169.0	0.420	1.262
Incoloy 825	1700° F-20 min WQ	66.7	138.0	283.0	0.353	0.715
Ni + 2% Be	1800° F Sol. T. WQ	41.0	104.0	222.0	0.39	1.00
Ni + 2% Be	As above + 1070° F-2 hr age	140.0	195.0	300.0	0.15	0.18
Ni +15.8Cr +7.2Fe	A	36.0	90.0	203.0	0.45	0.92
		Special Alloys				
Cobalt Alloy[8]	2250° F solution HT	65.0	129.0	300.0	0.50	0.51
As above	As above-trans[5]	65.0	129.0	300.0	0.50	0..40
Cobalt Alloy[7][9]	As rec'd (Ann'd)	62.8	119.5	283.0	0.52	0.75
As above	Machined + 2250° F sol HT	48.0	112.5	283.0	0.62	0.70
As above	2250° F sol HT +925° F aged	48.0	107.5	270.0	0.63	1.00
Molybdenum	Extr'd A	49.5	70.7	106.0	0.12	0.38
Vanadium	A	45.0	63.0	97.0	0.17	1.10

[7] 1/2" dia. bar.
[8] HS25 or L605 alloy. 50 Co 20 Cr 15 W 10 Ni 3 Fe.
[9] Elgiloy: 50 Co 20 Cr 15 Ni 7 Mo 15 Fe.

Table B-3. Tensile Properties at Low Strain Rates and Elevated Temperatures

Temp. °F	ε′ (per sec)	S_y (ksi)	S_u (ksi)	σ_o (ksi)	m	A_r (%)
\multicolumn{7}{c}{OFHC Copper (Annealed)}						
70	0.018	—	33.5	70.0	0.366	94
70	0.0024	—	32.5	67.5	0.361	92
70	0.0003	—	32.7	65.6	0.353	94
300	0.018	—	27.7	56.7	0.340	94
300	0.0024	—	26.7	53.8	0.334	91
300	0.0003	—	25.6	512.7	0.336	86
600	0.018	—	20.9	41.2	0.315	73
600	0.0024	—	20.4	38.0	0.268	62
600	0.0003	—	18.2	32.8	0.240	46
\multicolumn{7}{c}{70 Cu-30 Zn Brass (annealed)}						
70	0.018	13.8	48.5	120	0.606	80
70	0.0024	13.6	48.7	120	0.590	80
70	0.0003	13.6	48.5	120	0.568	77
200	0.018	14.8	45.2	108	0.536	75
200	0.0024	13.6	45.2	108	0.534	76
300	0.018	12.4	42.9	101	0.508	67
300	0.0024	13.6	43.4	101	0.507	61
300	0.0003	12.8	42.2	100	0.520	48
600	0.018	12.0	36.4	71.0	0.300	32
600	0.0024	12.5	33.2	58.5	0.225	25
600	0.0003	12.4	27.4	40.0	0.120	19
\multicolumn{7}{c}{60 Cu-40 Zn Brass (annealed)}						
70	0.018	14.2	53.5	134	0.620	74
70	0.0024	14.2	51.8	125	0.555	73
70	0.0003	13.2	50.0	122	0.600	68
300	0.018	13.9	43.6	102	0.505	66
300	0.0024	15.0	43.5	100	0.472	54
300	0.0003	14.8	39.0	93.0	0.520	54
600	0.018	16.0	27.0	46.0	0.200	23
600	0.0024	13.6	20.7	32.8	0.156	16
600	0.0003	11.6	16.0			16

Table B-4. Mechanical Properties of Some Plastics

Material	Strength Tensile (ksi)	Strength Compression (ksi)	Tensile Modulus (10^5 psi)	Elongation (%)	Rockwell Hardness	Izod (ft lb/in.)
ABS						
Medium impact	6-8	10-12	3-4	15-25	R_R 108-115	2-5
Very high impact	4-6	8-11	2-3	15-40	R_R 85-105	7-8
Acrylic						
Cast	6-12	11-19	3-5	5	R_M 80-102	0.5
Molding grade	9-11	12-19	3-5	5	R_M 85-95	0.4
Epoxy						
Cast rigid	9-15	15-35	4-5	5	R_M 106	
Molded	8-20	20-40	15-25	5	R_B 75-80	1.4
Glass cloth laminate	30-40	30-60	30-39	3	R_M 115-117	10-30
Filament-wound comp.	130-200	40-175	60-75	3	R_M 98-120	10-30
Fluorocarbon						
PTFE	2-7	1-2	1	350	R_J 79085	3-6
PVF	5-7	8-9	1	100-300	—	3-4
Nylons						
Nylon 6	9-12	6-13	4	150	R_R 118-120	1-6
Nylon 6/6	12	6-12	4-5	60-300	R_R 118	1-2
Phenolic						
Mineral/glass fiber	5-12	30-40	1-2F	1-2	R_E 50-90	1-9
Shock and heat	4-9	25-30	15-25	1-2	R_E 80-90	1.6
Polycarbonate	9-11	10-12	3-4	130	R_M 70	12-18
Polyester						
Cast rigid	6-13	13-36	9-12F	4	R_B 45-65	0.3
Polyethylene						
Low density	1-3	—	0.3	50-800	S_D 73	20
High density	3-4	3-4	1-2F	50/1000	S_D 63	1-5
Polypropylene	4-5	5-8	1.6	300	S_D 72	0.4-2
Polystyrene	5-8	11-16	4.6	1-3	R_M 72	0.6
PVC	1-4	1-2	0.03	350	S_A 50-100	—

Appendix C

Table C-1. Conversion from Cold Work to Strain

W (%)	Ul (%)	R (in.^{2}/in.^{2})	n (in./in.)	ε (in./in.)
0	0	1	0	0
5	5.2	1.052	0.052	0.051
10	11.1	1.111	0.111	0.105
20	25	1.250	0.250	0.223
25	33.3	1.333	0.333	0.288
30	42.8	1.428	0.428	0.355
33.3	50	1.500	0.500	0.405
40	66.7	1.667	0.667	0.511
50	100	2.000	1.000	0.693
60	150	2.500	1.500	0.916
63.2	172	2.718	1.718	1.000
70	233	3.333	2.333	1.203
75	300	4.000	3.000	1.386
80	400	5.000	4.000	1.609
90	900	10.000	9.000	2.303
95	1,900	20.000	19.000	2.996
98	4,900	50.000	49.000	3.912
99	9,900	100.000	99.000	4.605
99.9	99,900	1,000.000	999.000	6.908
99.995	2,202,500	22,026.000	22,025.000	10.000

W = cold work as percent reduction of area; Ul = uniform elongation; R = area ratio; n = nominal strain (valid only for uniform deformation); ε = natural strain.

Appendix C 347

Table C-2. Percent Cold Work (CW) vs. Natural Strain (ε)

CW	ε	CW	ε	CW	ε	CW	ε
1	0.010	26	0.301	51	0.714	76	1.427
2	0.020	27	0.315	52	0.743	77	1.470
3	0.030	28	0.328	53	0.755	78	1.514
4	0.041	29	0.342	54	0.777	79	1.561
5	0.051	30	0.355	55	0.799	80	1.610
6	0.062	31	0.371	56	0.821	81	1.661
7	0.073	32	0.385	57	0.844	82	1.715
8	0.083	33	0.400	58	0.868	83	1.772
9	0.094	34	0.416	59	0.892	84	1.838
10	0.105	35	0.431	60	0.917	85	1.897
11	0.115	36	0.446	61	0.942	86	1.966
12	0.128	37	0.462	62	0.968	87	2.041
13	0.139	38	0.478	63	0.995	88	2.121
14	0.151	39	0.494	64	1.022	89	2.208
15	0.163	40	0.511	65	1.050	90	2.302
16	0.178	41	0.528	66	1.079	91	2.408
17	0.183	42	0.545	67	1.109	92	2.526
18	0.200	43	0.562	68	1.140	93	2.660
19	0.211	44	0.580	69	1.172	94	2.814
20	0.223	45	0.598	70	1.204	95	2.886
21	0.236	46	0.616	71	1.204	96	3.219
22	0.248	47	0.635	72	1.273	97	1.109
23	0.261	48	0.654	73	1.310	98	3.912
24	0.274	49	0./673	74	1.347	99	4.605
25	0.288	50	0.693	75	1.387	99.99	10.000

Appendix D

Table D-1. Designations for Aluminum Alloy

Type	Old number	New number	Suffix	Meaning
Aluminum (99.00% min)	29	1XXX	O	Annealed
Alloys (grouped by major alloy)				
Copper	11s, 14s, 17s, 24s	2XXX	H	Strain-hardened
Manganese	3s	3XXX	F	As fabricated
Silicon	32s	4XXX	W	Solution treated
Manganese	52s	5XXX	T	Heat treat temper (T2 to T10 for wrought materials)
Magnesium and silicon	61s	6XXX		
Zinc	75s	7XXX		

Subdivisions of —T temper

—T2 Annealed (cast product only)
—T3 Solution heat treated and cold worked
—T4 Solution heat treated and naturally aged
—T5 Artificially aged
—T6 Solution heat treated and artificially aged
—T7 Solution heat treated and stabilized
—T8 Solution heat treated, cold worked, and artificially aged
—T9 Solution heat treated, artificially aged, and cold worked
—T10 Artificially aged and cold worked

Table D-2. Chemical Composition of Some Aluminum Alloys

Number	Principal alloy element	Alloy elements				
		Cu	Mn	Si	Mg	Zn
1100	None	—	—	—	—	—
2011	Copper	5.5	—	—	—	—
2014	Copper	4.5	0.8	0.8	0.4	—
2017	Copper	4.0	0.5	—	0.5	—
2024	Copper	4.0	0.6	—	1.5	—
3003	Manganese	—	1.2	—	—	—
4032	Silicon	0.9	—	12.2	1.1	—
5052	Magnesium	—	—	—	2.5	—
6061	Magnesium	0.25	—	0.6	1.0	—
7075	Zinc	1.6	—	—	2.5	5.6

Table D-3. Typical Mechanical Properties of Wrought Aluminum Alloy

Alloy and temper	Strength (psi)		Percent Elongation	BHN-500 kg
	Tensile	Yield		
1100-O	13,000	5,000	45	23
1100-H18	24,000	22,000	15	44
2011-T8	59,000	45,000	12	100
2014-O	27,000	14,000	18	45
2014-T6	70,000	60,000	13	135
2017-O	26,000	10,000	22	45
2017-T4	62,000	40,000	22	105
2024-O	27,000	11,000	22	47
2024-T3	70,000	50,000	18	120
4032-T6	55,000	46,000	9	120
6061-T6	45,000	40,000	17	95
7075-T6	82,000	72,000	11	150

Appendix E

Table E-1. AISI and SAE Designation of Structural Steels

The common steels are specified by a code consisting of four characters as shown below. A steel specification XYZZ has the following meaning:

X — indicates the type of alloy present
Y — indicates the percent of principal alloys
ZZ — indicates the "points" of carbon (points of carbon are % × 100)

When X Is	Alloys Present Are
1^1	None (plain carbon)
2	Nickel
3	Nickel-chromium
4	Molybdenum (+ nickel-chromium)
5	Chromium
6	Chromium-vanadium
8	Nickel-chromium-molybdenum
9	Silicon-manganese

Examples

2130 is a 1% nickel steel with 0.3% C
4340 is a 3% (molybdenum + Ni + Cr) steel with 0.4% C

---

[1] 11ZZ and 12ZZ indicate resulfurized free machining steel. 13ZZ indicates high-manganese, plain carbon steel. 15ZZ indicates medium manganese, plain carbon steel.

Table E-2. Composition of Some AISI Steels

Carbon Steels		
1% Mn max; 0.15/0.30 Si; 0.04 P max; 0.05 S max		
AISI/SAE	C	Mn
1020	.18/.23	.30/.60
1045	.43/.50	.60/.90
1045H	.42/.51	.50/1.00
1080	.75/.88	.50/.90

Resulfurized: Free Machining			
AISI/SAE	C	Mn	S
1117	.14/.20	1.00/1.30	.08/.13
1144	.40/.48	1.35/1.65	.24/.33
1212	.13 max	.70/1.00	.16/.23

Carbon Steels (Higher Mn)*		
AISI/SAE	C	Mn
1522	.18/.24	1.10/1.40
1566	.60/.71	.85/1.15
1340	.38/.43	1.60/1.90

Alloy Steels					
AISI/SAE	C	Mn	Ni	Cr	Mo
4027	.25/.30	.70/.90	---	---	---
4820	.18/.23	.50/.70	3.25/3.75	---	.20/.30
52100	.98/1.10	.25/.35	---	1.30/1.60	---
9260[2]	.56/.64	.75/1.00	---		---
50B40	.38/.43	.75/1.00		.40/.60	

*Note: 15ZZ steel has 1.00-1.50% Mn
13ZZ steel has 1.60 - 1.90% Mn
XYBZZ steel has 0.0005 - 0.002% B

---

[2]Contains 1.80 to 2.20% Si

Appendix F

Table F-1. Approximate Prices of Castings[1]

Sand Castings		Die Castings	
Material	Cost $/lb	Material	Cost $/lb
Gray iron	0.35-0.65	Zinc base	1.25-2.00
Malleable iron	0.55-0.75	Aluminum	2.00-2.50
Steel	0.70-0.95	Brass	2.00-3.50
Aluminum	2.00-2.50	Plastic (Nylon)	2.00-4.50
Magnesium	3.00-5.00	Permanent mold iron	0.40-0.85
Brass	2.00-3.50		

[1] The lower cost figure will usually apply when the quantity is very large and the shape is very simple. The higher cost figure will usually apply when the quantity is small and the shape of the casting is complex.

Table F-2. Warehouse Prices of Steel

Material	Cost $/lb	Material	Cost $/lb
Low carbon steel		Alloy steel bars	
HR sheet (18 ga.)	0.45	HR 4615 (2 in. dia.)	0.72
CR sheet (15 ga.)	0.56	CR 4615 (2 in. dia.)	0.82
HR strip	0.55	HR 4140 (2 in. dia.)	0.69
HR plate	0.47	CD 4140 ann (2 in. dia.)	0.79
HR structural shapes	0.35		
HR bars (2 in. dia.)	0.38		
CF bars (2 in. dia.)	0.48		

Table F-3. Market Prices of bulk plastics
(1996 prices, $/lb. Price range depends upon the grade.)

Resin	Price	Resin	Price
ABS	0.82-1.99	Nylon	1.22-3.50
Acetol	1.23-1.50	Polyamide	19.75-22.25
Acrylic	1.00-1.73	Polycarbonate	1.37-2.50
Epoxy	1.16-2.70	Polyester	0.57-1.50
Teflon	11.00-20.00	Polystyrene	0.45-0.94

Appendix G

Forming Operations Performed on Presses and Hammers

Operation	Definition or Description	Sketch	Uses or Applications
Open die forging	Forming by uniaxial compression between flat parallel plates (dies)		Discs or blanks used as preforms. Very large shafts or forgings of simple shape
Closed die forging	Forming by a uniaxial compressive force using dies with cavities which control lateral flow		Shaped parts small to medium size such as connecting rods, handles, levers, gear blanks
Upsetting	A special type of closed die forging operation on bars in which a longitudinal force causes a local flow		Heads on bolts, valves. Flanges and shoulders on shafts
Forward extrusion	Forming by axial flow of a billet through a die orifice in the direction of an applied compressive force		Regular or irregular structural shapes such as window moldings, angle sections, circular, or irregular tubing
Back extrusion	Forming by axial flow of a billet through a die orifice opposite to the direction of an applied compressive force		Regular or irregular structural shapes, tubes for grease and cosmetics
Coining	Special type of closed die forging operation in which the lateral surfaces are restrained resulting in a variable thickness and a well-defined imprint of the die faces		Shallow configurations on flat objects, such as ornamental designs or structural ribs and bosses

Appendix G

Operation	Definition or Description	Sketch	Uses or Applications
Hobbing (Hubbing)	Forming a very smooth, accurately shaped die cavity by pressing a hardened punch into a softer metal die block		Making many duplicate cavities, e.g. plastic molds for molding typewriter keys, plastic wall tile
Embossing	Forming a design on thin materials by raising or lowering portions of the material by localized bending		Forming ribs to increase rigidity. Making ornamental or communicative designs
Drawing (Deep, Cup)	Forming by pushing a formed punch against a flat sheet and forcing it over a shaped die-edge into the die cavity to take the shape of a recessed vessel.		Cups, shells short tubes, automotive bodies, gas tanks, appliance covers
Ironing	Forming by reducing the wall thickness of deep drawn parts by forcing them through a die with a punch wherein the clearance is less than the original wall thickness		For thinning and sizing walls of predrawn parts
Sizing	Finish forming operation, usually done on a preformed part, in closed dies to obtain very accurate dimensions		Accurate control of height or area of a formed part such as bosses

Appendix G   355

Operation	Definition or Description	Sketch	Uses or Applications
Braking (Bending)	Forming of structural shapes such as angles and channels by bending sheets or plates in a long press called a brake		Structural shapes such as automobile and truck chassis. Angle and channel sections having large moment of inertia to weight ratio
Hydroform (Guerin Process) (Rubber Punch)	Bending or drawing operation in which either the punch or the die is a piece of constrained rubber		Low quantity of parts that could be made by drawing
Swagging	Forming operation in which the diameter of bars or tubes is changed by repeated blows of radially actuated shaped hammers		Flaring ends of tubes, reduction of tube or shaft diameter
Staking	Forcing a shaped punch into the top of a projection to complete a riveting operation		As a substitute to drilling and riveting
Shearing	A material separation operation using two knife blades or a punch and die wherein the material deformation is localized and due to shear forces		Rapid method of cutting bar stock or plates to length

356   Appendix G

Operation	Definition or Description	Sketch	Uses or Applications
Blanking	A material separation process using a punch and die to shear a useful blank from a thin piece of material		Making discs, washers or flat blanks which may be drawn afterwards
Piercing	A material separation operation using a punch and die to shear a hole in a thin piece		Rapid method of making holes or slots in a part

Forming Operations Performed on Rolls

Operation	Definition or Description	Sketch	Uses or Applications
Rolling	Forming operation on cylindrical rolls wherein the cross sectional area of a bar or plate is reduced with a corresponding increase in length		Rapid method of producing bars of uniform cross section
Roll forming	A continuous bending operation generally performed on originally coiled strip material and resulting in long shaped sections such as angles, channels and tubes		Rapid method of forming long lengths of tubes or structural shapes
Forge rolling	A closed die forging operation in which the die cavities are machined in the periphery of cylinders		Similar to closed die forging except longer parts can be forged

Appendix G

Operation	Definition or Description	Sketch	Uses or Applications
Power spinning	Forming of a conical part from a disc in which the diameter does not change by progressively forcing the disc against a conical mandrel by means of a small roller moving from the center out		For low production forming since cost of tooling is low
Roll straightening	Straightening of a bar or plate by passing it through a series of rolls which are not all in the same plane, causing the material to be bent so that it is straight after the bending forces are removed		To straighten sheets that have been rolled
Tube piercing	Tube piercing is a rolling process in which a solid cylindrical bar is formed into a rough seamless tube by means of two conical rolls oppositely skewed a few degrees from the axis of the bar		To originate holes in seamless tubing
Thread rolling	A plastic deformation process in which threads are formed on a cylinder that has an original diameter equal to the final pitch diameter		To form threads by plastic flow rather than machining. Gear teeth and splines can be formed by a similar process

358 Appendix G

Forming Operations Performed on Draw Benches

Operation	Definition or Description	Sketch	Uses or Applications
Wire or bar drawing	Reducing the diameter of a bar or wire by pulling it through a tapered orifice in a die	d_1, d_2, F; $d_2 < d_1$	Forming small cross sectional parts of uniform cross section and close tolerance
Tube drawing	Reducing the wall thickness of a tube by pulling it simultaneously over a mandrel and through a die orifice	Die, Tube, Mandrel, F	Forming close tolerance tubing
Stretch forming	Simultaneously stretching and bending, over a die, a bar or plate that would fail by compressive buckling if bending alone was attempted	F, F	Bending "I" or "L" sections to small radii
Stretcher levelling	Straightening a sheet or plate by performing slight tensile deformation	F, F	Straightening sheets or plates previously rolled

Appendix H

The Machining Operations

Machining is the manufacturing process in which the size, shape, or surface properties of a part are intentionally changed by removing the excess material.

A. Mechanical Machining: The process of removing the excess material by means of a cutting tool which locally strains the material to fracture.

1. Operations Done on Turning Machines.

Operation	Definition	Sketch	Tolerances[1] (±, in.)	Surface[2] Finish (rms)	Remarks
Turning	Turning is the machining operation for generating external surfaces of revolution on a rotating work piece by feeding a tool in a direction that is not perpendicular to the axis of rotation		0.005 0.001	100 - 2000 16 - 200	The tool feeds parallel to the axis of rotation for cylindrical surfaces and at an angle to the axis of rotation for conical or irregular surfaces. Light feeds are .0005 to .005 i.p.r.; heavy feeds are .010 to .060 i.p.r.
Form Cutting	Form cutting is the machining operation for producing surfaces of revolution on a rotating work piece whose cross sections are reciprocal profiles of the cutting edge of the "form tool" used.		0.005 0.001	60 - 500 16 - 200	The form tool may be fed radially to form features such as grooves or shoulders on shafts; or else the tool may be fed parallel to the axis of rotation to form recesses or projections on the face of the work piece. The feeds are usually less than for turning.
Boring	Boring is the machining operation in which internal surfaces of revolution are generated on a rotating work piece by a feeding motion that is not perpendicular to the axis of rotation		0.005 0.001	60 - 1000 8 - 150	Boring is a hole enlarging operation and not a hole originating operation. The cutting is done by a small single-point tool attached to the end of a cantilevered bar. Feeds are usually lower than for turning.

[1] The top number is a good, liberal, easy-to-obtain tolerance and the bottom number is the smallest tolerance that can be obtained repetitively without the need of filing or sanding.

[2] The top range is for roughing cuts and the bottom range is for finishing cuts. The range is affected by the feed and nose radius.

360   Appendix H

Operation	Definition	Sketch	Tolerances[1] (±, in.)	Surface[2] Finish (rms)	Remarks
Facing	Facing is the machining operation for generating plane surfaces by a cutting tool, with both the machine surface and the tool feed being perpendicular to the axis of a rotating workpiece.		0.005 0.001	100-2000 16-200	Since the spindle RPM does not vary during a facing operation, the cutting speed varies as the tool moves radially.
Drilling, Reaming	See operations done on Drilling Machines.				
Threading	Threading is the machining operation for producing threads on a rotating work piece by feeding, parallel to the center-line of the spindle, a single-point cutting tool having the shape of the thread profile				Other operations for producing threaded surfaces are die threading, thread rolling, and thread grinding. Where speed and production are important, die-cut threads are preferred over single point threads.

2. Operations Done On Milling Machines

Operation	Definition	Sketch	Tolerances[1] (±, in.)	Surface[2] Finish (rms)	Remarks
Face Milling	Face Milling is the machining operation which produces flat surfaces by means of a shank-mounted, disc-shaped cutter whose axis is perpendicular to both the machined face and the feeding direction of the work piece.		0.005 0.001	60 - 2000 16 - 100	Face Milling is an efficient method of removing metal. Feeds of 0.001 to 0.020 ipt can be used. These are somewhat larger than those used for side or slab milling.
End Milling	End Milling is the machining operation which produces surfaces by means of a cylindrically-shaped cutter having teeth on the end as well as on the periphery, and its axis of rotation may be either parallel or perpendicular to the machined face and the direction of feeding.		0.005 0.001	60 - 2000 16 - 100	End Milling is very similar to face milling with a small diameter cutter. Due to the larger beinding stresses, the feeds for end milling are lower than face milling.

Appendix H   361

Operation	Definition	Sketch	Tolerances[1] (±, in.)	Surface Finish (rms)	Remarks
Slab Milling	Slab Milling is the machining operation where a surface is produced by a multiple tooth cutter having a width to diameter ratio greater than one, and its axis of rotation perpendicular to the feeding motion and parallel to the machined face of the work piece.		0.005 0.001	60 - 2000 16 - 100	Chatter is a problem with many slab milling operations since the cutting edge is long. Feeds are generally lower than for side milling or face milling.
Side Milling	Side Milling is the machining operation where a surface is produced by a multiple tooth disc-shaped, arbor-mounted cutter having a width to diameter ratio less than one, its axis of rotation perpendicular to the feeding motion and either parallel or perpendicular to the machined face of the workpiece.		0.005 0.001	60 - 2000 16 - 400	Side milling can be used as a slotting operation. Feeds may vary from 0.0005 ipt for thin, large diameter cutters to 0.010 ipt for rigid cutters and soft work material.
Straddle Milling	Straddle Milling is the machining operation where two parallel faces are simultaneously cut with two parallel side milling cutters separated by a space equal to the distance between the two surfaces cut.		0.005 0.001	60 - 2000 16 - 100	Straddle milling is simply milling with two or more side milling cutters simultaneously. Feeds are similar to those for side milling.

3. Operations Done on Drilling Machines

Drilling	Drilling is the machining operation which originates or enlarges a hole with an end-cutting tool having one or more cutting edges. The cutting is done by rotating either the tool or the workpiece about the centerline of the hole.		See Remarks	60 - 1000	Drilling is a widely used operation requiring inexpensive tools. The tolerances on the diameter of the hole vary from +0.000 to +0.005" for drills up to 1/2" diameter and from +0.001 to 0.020 for drills up to 2" diameter.

Appendix H

Process	Definition	Sketch	Tolerances (± in)	Surface Finish (rms)	Remarks
Reaming	Reaming is the machining operation for enlarging an existing hole with a multiple-tooth end cutting tool to improve the accuracy and surface finish of a round hole.		0.002 0.0005	16 - 125	Reaming does not improve the accuracy of the location or angular alignment of a drilled hole but it gives a better surface finish and tolerance on the diameter. Large feeds are used.
Counter- sinking	Counter sinking is the machining operation where a conical end-cutting tool is used to cut a tapered enlargement at the opening of a hole for recessing the head of a fastener.			16 - 250	Recommended speed for counter sinking is normally one-half to two-thirds the speed used for drilling the same material.
Counter- boring	Counter boring is the machining operation for enlarging the diameter of a hole for a portion of its depth.		0.005 0.001	16 - 250	Counter boring is done by means of a multiple-tooth, end-cutting tool called a counterbore. It may provide a recess for a bolt head or a nut.
Spot- facing	Spot facing is the machining operation for producing a flat seat for a bolt head, washer, nut or similar element at the opening of a drilled hole, concentric with the hole and at right angles to its longitudinal axis.			16 - 250	Spot facing sometimes preceeds drilling to provide a contoured workpiece with a flat surface so as to facilitate centering and starting a drilled hole.
Tapping	Tapping is the machining operation for producing internal threads by a cylindrical or conical thread-cutting tool, called a tap, in a previously formed hole.		—	—	A tap resembles a screw that has either straight or helical flutes machined in it which provide the cutting edges for the tap.

Appendix H

4. Operation Done on Shaping Machines

	Definition	Sketch	Tolerances[1] (±,in.)	Surface[2] Finish (rms)	Remarks
Shaping	Shaping is the machining operation for removing metal from surfaces in horizontal, vertical or angular planes, by the use of a single-point tool supported by a ram that reciprocates the tool in a linear motion.		.010 .002	63-2000	A special type of shaping process which uses a vertical ram with its tool reciprocating vertically is called sloting.
Planing	Planing is the machining operation for removing metal from surfaces in horizontal, vertical or angular planes, in which the workpiece is reciprocated in linear motion past one or more single-point tools.		.010 .002	32-2000	Suited primarily for producing flat surfaces on parts too long and too cumbersome for the shaper.

5. Operations Done on Miscellaneous Machines

	Definition	Sketch	Tolerances[1] (±,in.)	Surface[2] Finish (rms)	Remarks
Broaching	Broaching is the machining operation in which a broach is pulled or pushed over the face of a workpiece, or through a previously created hole in the workpiece, to remove metal by axial cutting.		.005 .0005	16-250	A broach is similar to a hacksaw blade in that it consists of a row of uniformly spaced cutting teeth. Each tooth extends above the centerline of the broach slightly beyond that of the preceeding tooth.
Sawing	Sawing is the machining operation in which material is cut by the action of a thin multiple-tooth tool in the form of either a blade, band, or a disc.		.020 .005	125-2000	Friction and abrasive sawing are two types of sawing performed with high speeds, resulting in enough frictional heat to melt the material. Friction sawing is done on a band saw while abrasive uses a thin grinding wheel.

364 Appendix H

B. Non-Mechanical Machining: The process of removing excess material by chemical or electrochemical techniques or by the melting or vaporizing effects of a high energy source.

Operation	Definition	Sketch	Tolerances[1] ($\pm$, in.)	Surface[2] Finish (rms)	Remarks
Chemical Machining	Chemical Machining is the process where material is removed from a part by putting it in contact with a proper reagent so that the material is removed by means of a chemical reaction.		.005 .001	15 - 250	This process is particularly useful in maching large, complex shapes, sections too thin to tolerate the stress of conventional machining, and in removing metal from exceptionally hard or brittle materials.
Flame Cutting	Flame Cutting is a chemical machining process where the material is removed by oxidation with a fine jet of oxygen, according to the reaction $xM + yO \rightarrow M_xO_y$.		.06 .02	100-2000	This process uses the same equipment used for welding operations.
Electrochemical Machining	Electrochemical Machining is the process of material removal by dissolution in an electrolytic cell in which the workpiece is the anode and the tool is the cathode.		.005 .0005	100 - 2000	
Electric Discharge Machining	Electric Discharge Machining is the process of removing material by melting or vaporization using the highly concentrated energy of an electric discharge to focus a high temperature on a small area of an electrically conductive workpiece.		.005 .001	30 -500	In this process the surface properties may change because of high temperatures and rapid cooling. It is a slow process, uneconomical for most machining of soft materials.

Appendix H 365

Operation	Definition	Sketch	Tolerances[1] (±, in.)	Surface[2] Finish (rms)	Remarks
Electron Beam Machining	Electron Beam Machining is the process of cutting materials with a focused beam of high velocity electrons which convert their kinetic energy to heat on impact with the workpiece and cause it to melt or vaporize.		.001 .0005	10-100	This process is used in micro-machining operations, usually on hard materials. Its biggest disadvantage is that the work must be performed in a vacuum.
Laser	Laser machining is the cutting process in which the work material is melted and vaporized by a narrow beam of intense monochromatic light.		.001 .0005	10-100	Laser machining is used only in microdrilling and microwelding operations due to its low power output. Unlike the electron beam, a laser need not operate in a vacuum.
Ultrasonic Machining	Ultrasonic Machining is the process of material removal by micro-fracture induced by the impact of abrasive particles on the work-piece.		.003 .0005	5-64	The abrasive particles vibrate in a liquid slurry circulating through a narrow gap between the workpiece and a tool oscillating at high frequency.

366 Appendix H

6. Operations Done on Grinding Machines

	Definition	Sketch	Tolerances[1] (± in.)	Surface Finish[2] (rms)	Remarks
Abrasive Wheel Grinding	Abrasive wheel grinding is the machining operation in which abrasive particles bonded together as a disc, are used to chip small amounts of material from the workpiece.		.001 .0002	8 - 125	The feed is obtained by moving the workpiece past the high speed (4,000 to 10,000 fpm) grinding wheel. Some types of grinding are: surface, cylindrical and centerless grinding.
Abrasive Belt Grinding	Abrasive belt grinding is the machining operation in which an endless belt precoated with abrasive particles is used to remove stock from the surface of the workpiece.		.001 .0002	4 - 125	Contact wheels, over which the belt rides, directly affect the rate of stock removal and surface finish.
Honing	Honing is the grinding operation in which small abrasive grains of a "honing stick" are used to remove small amounts of stock from the workpiece.		.0005 .000,05	1 - 63	Honing is most often used on interior cylindrical surfaces in a simultaneous rotating and reciprocating action of the honing tool. It produces smooth, precise surfaces.
Lapping	Lapping is the grinding operation in which very small, loose, abrasive grains are rubbed at low speed and low pressure aginst the workpiece.		.0002 .000,05	0-16	Lapping results in: a) extreme accuracy, b) correction of minor imperfections of shape, and c) refinement of surface finish, and extremely close fit between mating surfaces.

Index

ABS, 171
Acetylene welding, 290
Acicular, 122, 124
Acrylic, 171
Activation energy, 114
Adams, C. M., 298
AISI specification, 157, 350
Allotropism, 72
Alloys
    interstitial, 54
    multiple phase, 66
    single phase, 65
    substitutional, 77
Alpha iron (also α iron), 74
Aluminum, 180, 348
    composition, 349
    designation, 348
    strength, 349
    temper, 348
Aluminum-copper phase diagram, 84

Amount of phases, 79
Arc length, 296
Arc welding, 290
    bead size, 294
    burn off rate, 292
    HAZ, 295
    load capacity, 295
    MIG, 290
    penetration, 294
    stick, 290
    submerged arc, 290
    TIG, 290
Area ratio, 203
Aromatic hydrocarbons, 37
Arrhenius rate equation, 117
Atomic
    bonds, 25
        covalent, 28
        ionic, 26
        metallic, 30

368 Index

[Atomic]
    molecular, 32
    van der Waals, 32
  number, 26
  packing factor, 46
  radius, 46
  structure, 33
    amorphous, 34
    crystalline, 42
    molecular, 34
  weight, 26
Atomic bonds, 25
Atomic number, 26
Atomic packing factor, 46
Atomic radius, 46
Atomic structure, 33
Atomic weight, 26
Austempering, 132
Austenite (also γ iron), 74, 104
  retained, 123
Austenite transformation rate, 126
Axial deformation, 261
  compressive, 261
  tensile, 261

Back-rake angle, 317
Bainite, 124, 128
  mechanical properties, 125
Bakelite, 40
Bead size, 293
Bending deformation, 261
  analysis, 263
  strains, 261
  strain history, 263
Binary phase diagrams, 76
  application rules, 78
  aluminum-copper, 84
  copper-tin, 86
  copper-zinc, 83
  iron-carbon, 85
  iron-lead, 87
Bisacre, G. W., 328

Blue brittleness, 145
Body-centered cubic, 44
Bordon, M. P., 256
Boron, 157
Boston, O. W., 330
Boundaries, 54
Brass, 83
Breaking strength, 202
Brine, 161
Brinell hardness, 180
Burgers vector, 57
Burn off rate, 293
Butane, 36

Carbide, 314
Carbon content of pearlite, 106
Cast iron, 85
Cast non-ferrous, 314
Cementite, 74, 104
Ceramic, 314
Chao, B. T., 328
Coarse pearlite, 121
Cobalt, 343
Cold pressure welding, 290
Cold rolling, 261
Cold work, 64, 194
Composition of phases, 78
Compressive strain, 193
Continuous transformation diagram, 133
Conversion factors, 339
Cooling curves, 71
  of 1080 steel, 133
Coordination number, 45
Copper alloys, 170
Copper-nickel phase diagram, 76
Copper-tin phase diagram, 87
Copper-zinc phase diagram, 83
Cored structure, 79
Cost effective, 155
Cost of
  alloy steel, 166, 168

[Cost of]
   aluminum alloys, 170
   carbon steel, 166, 167
   copper alloys, 170
   polymers, 171
Coulomb force, 26
Covalent bond, 26
Critical cooling rate, 134
Critical diameter, ideal, 161
Crystal
   directions, 50
   imperfections, 53
      dislocations, 53
      point defects, 54
      planar defects, 57
   planes, 50
   structures, 45
      body-centered cubic, 47
      face-centered cubic, 45
      hexagonal close packed, 49
Crystal direction, 50
Crystal imperfections, 53
Crystal planes, 50
CT diagram, 135
Cutting condition, 308
   cutting speed, 309
   depth of cut, 309
   feed, 309
Cutting speed, 309
Cutting tools, 308
   cast non-ferrous, 314
   ceramic, 314
   high-speed steel, 314
   sintered carbide, 314
Cycle of strain, 256

Datsko, J., 7, 250, 325
Deformation
   elastic, 60, 189, 192
   energy, 272
   force, 272
   plastic, 61, 192

[Deformation]
   strengthening, 247
   ulastic, 210
   uniform, 205
   work, 272
Deformation modes, 261
   axial, 261
   bending, 261
   rolling, 261
   shear, 267
   torsion, 267
Deformation strain, 260
Density, 340
Depth of cut, 309
Design rules, 7
Diamond pyramid hardness, 184
Diffusion, 107, 114, 124
   coefficient, 116
   grain boundary, 115
   interdiffusion, 115
   surface, 116
   volume, 115
Diffusion coefficient, 116
Diffusivity, 116
Dipoles, 33
Direction indices, 50
Dislocation, 53
   edge, 56
   movements, 60
   screw, 56
Dispersion hardening, 66
Drill press, 311
Dynamic hardness, 187

Edge dislocation, 56
Effective rake angle, 332
Elastic deformation, 60, 189, 192
Elastic-plastic region, 198
Electron beam welding, 290
End-cutting angle, 317
End-quench hardenability, 157
   limits, 159, 160

Index

Energy absorption, 273
Energy of deformation, 272
Engineering stress-strain, 188
Equicohesive temperature, 64
Equilibrium phase diagrams, 69
    binary, 76
        application rules, 78
        Al-Cu, 84
        Cu-Sn, 86
        Cu-Zn, 83
        Fe-C, 85
        Fe-Pb, 87
    ternary, 69
    unary, 69
Equilibrium microconstituents
    alpha, 96
    austenite, 104
    betta, 96
    cementite, 99
    eutectic, 96
    eutectoid, 99
    pearlite, 99
Equilibrium phases
    alpha, 96
    austenite, 104
    betta, 96
    cementite, 99
Equilibrium transformation, 120
Equivalent energy input, 297
Equivalent strain, 256
Ethylene, 38
Eutectic, 73
    composition, 73, 98
    temperature, 73
Eutectoid, 97
    composition, 104
    reaction, 74
Extensometer, 192

Face-centered cubic, 44
Fe-C microconstituent diagram, 99
Fe-C phase diagram, 86

Feed, 309
Feed rate, 310
Ferrite, 74, 104
Ferromagnetic, 72
Ferrous metals, 155, 350
Ferrous microconstituents, 103
    austenite, 104
    cementite, 104
    ferrite, 104
    pearlite, 104
Fick's law, 116
Fillet welds, 300
    load capacity, 301
    size, 300
Fine pearlite, 121
Force of deformation, 273
Forge welding, 290
Formability, 247
Fracture strain, 203
Fracture strength, 202
Frenkel defect, 56
Friction welding, 290
Fundamental machinability, 324
Fusion, 287

Gage length, 187
Gage section, 187
γ-to-α transformation, 109
γ-to-pearlite transformation, 109
Gas constant R, 117
Gas welding, 290
Gibbs, W., 77
Glass transition temperature, 33
Grain, 63
    boundary, 56, 115
    size, 64
    strength, 64
Grain size, 64
    multiplying factors, 163
Grinding machine, 312
Grossman, M. A., 161
Growth, 107

Hardenability, 136
 and composition, 161
 band, 139
 Jominy curves, 137
 limits, 157
 precipitation, 147
Hardness, 178
 Brinell, 180
 coefficient, 182
 diamond pyramid, 184
 dynamic, 187
 exponent, 182
 indenters, 179
 Knoop, 185
 Meyer, 181
 microhardness, 186
 Rockwell, 178
 scleroscope, 186
 static, 187
Heat-affected zone, 295
Heat capacity, 323
Heat of fusion, 340
Heating curves, 71
Heat utilization, 300
Henkin, A., 325
Herbert, E. G., 325
Hexagonal close-packed, 44
High-speed steel, 314
Hill, R., 262
Hoffman, D., 262
Hooke's law, 195, 200
H values, 163
Hypereutectoid, 129
 steel, 102
Hypoeutectoid steel, 102, 129

Ideal critical diameter, 161
Ideal material, 198
Ideal quench, 161
Identity of phases, 78
Impact strength, 144
Imperfection, 53

Inconel, 343
Indenters (hardness), 179
Induction welding, 290
Infinite quench, 161
Inflection point, 75
Interdiffusion, 115
Intermetallic compound, 84
Interstitial atom, 54
Interstitial defect, 55
Inverse lever rule, 79
Ion, 26
Ionic bond, 26
Interrupted quench, 131
 austempering, 132
 martempering, 132
Iron, 61
Iron carbide (Fe_3C), 74
Iron-carbon phase diagram, 86
Iron-carbon microconstituent diagram, 100
Iron-lead phase diagram, 88
Isomers, 36
Isothermal phase change, 72
Isothermal transformation, 72
 austempering, 132
 diagrams, 125
 martempering, 132
Isothermal transformation diagrams, 125
 AISI 1040 steel, 129
 AISI 1080 steel, 124
 AISI 4340 steel, 130
 Atlas of, 125
 C curves, 125
 S curves, 125
 TTT, 125
IT diagram, 125

Jhaveri, P., 298
Jog, 62
Johnson, W., 262

Jominy, 137
  apparatus, 138
  cooling rate, 138, 140
  curves, 137
  distance, 138
  end-quench test, 138
  equivalent distance, 140
  test, 138

K_b values, 214
Knoop hardness, 185

Lamellar pearlite, 99
Lamellar structure, 97
Laser welding, 290
Lathe, 310
Lattice parameter, 110
Lead-tin microconstituent diagram, 97
Lead-tin phase diagram, 81
Ledeburite, 99
Line defects, 53
Load capacity, 295
Lower critical temperature, 101
Lower yield point, 202

Machinability, 320
  fundamental equation, 324
Machine tool, 308
  drill press, 311
  grinding machine, 312
  lathe, 310
  milling machine, 312
  shaper, 312
Machining, 308
Macromolecules, 39
Martempering, 132
Martensite, 121
  primary, 121
  reaction, 123
  secondary, 122
  tempered, 122

Martensite reaction, 123
Mass production, 307
Mechanical properties
  and carbon content, 158
  ferrous microconstituents, 156
  interrelationships, 177
  molecular structures, 41
  rules of, 101
  science of, 2
  structure-insensitive, 177
  structure-sensitive, 177
Mechanical reliability, 5
Mechanical strength, 63
  principles of, 63
    cold work, 64
    grain size, 64
    multiple phase alloying, 65
    single phase alloying, 65
Medium pearlite, 121
Mellor, P. B., 262
Meyer, 181
  hardness, 181
  strain hardening, 182
Metallic bond, 26
Metallic compound, 84
Metallurgical microscope, 92
Metal removal rate, 310
Methane, 35
M_f temperature, 124
Microconstituent, 92
  diagrams, 96
  Fe-C, 99
  ferrous equilibrium, 103
  Pb-Sn, 96
Microhardness, 186
Microscopy, 92
Microstructure, 92, 95
Miller indices, 51
Milling machine, 312
Mitchell, W. J., 209, 211
Modulus of elasticity, 60, 103
Moffatt, W. G., 298

Molecular bond, 26
Molecule, 34
Molybdenum, 343
Monel, 343
Monomers, 37
M_s temperature, 124
Multiple phase alloy, 66
Multiplying factors, 161
    for alloys, 161, 163
    for carbon, 163
    for grain size, 163

NaCl lattice, 28
Necking, 191
Neutral axis, 261
Neutral surface, 262
Neutrons, 25
Nickel, 343
Nominal strain, 189
Nominal stress-strain, 188
Non-equilibrium diagrams, 113
Non-equilibrium microstructure, 113
    Bainite, 124
    fine pearlite, 121
    martensite, 121
        primary, 121
        secondary, 122
        tempered, 122
    medium pearlite, 121
    retained austenite, 123
Non-destructive test, 178
Non-equilibrium phases, 121
    Bainite, 124
    martensite, 121
    retained austenite, 123
Non-isothermal phase change, 75
Nose radius, 320
Nucleation, 107, 124
Nucleus, 25
Nylon, 171

Offset strain, 201

Oil quench, 140
Olefins, 37
Optimal material selection, 3
Optimal properties, 91, 96, 161
Order-disorder reaction, 83
Oriented polymers, 42
Over yield, 198

Paraffin, 35
Pb-Sn microconstituent diagram, 97
Pearlite, 99, 104
    fine, 121
    medium, 121
    reaction, 106
    spheroidized, 99
Pearlite nose, 129
Pearlite reaction, 106
Penetration, 287, 294
Percent elongation, 203
Plastic deformation, 192
Peritectic reaction, 75
Peritectoid, 75, 83
    reaction, 75
Phase, 70
    amount of, 79
    change, 71
        isothermal, 71
        non-isothermal, 75
        rules of, 72
        transformation, 73
    composition of, 78
    diagrams, 76
        application rules, 78
        binary, 69
        ternary, 69
        unary, 69
    identity of, 78
Phase diagrams, 69, 76
    Al-Cu, 84
    Cu-Ni, 76
    Cu-Sn, 87
    Cu-Zn, 83

374 Index

[Phase diagrams]
   Fe-C, 86
   Fe-Pb, 88
   Pb-Sn, 81
Phenal formaldehyde, 40
Planar defects, 54
   grain boundary, 57
   low angle, 59
   stacking fault, 59
   twin boundary, 58
Plastic deformation, 60, 61
Plexiglass, 171
Point defects, 53
Poisson's ratio, 60
Polymerization, 37
   thermoplastic, 40
   thermosetting, 40
Polymers, 171
Polymide, 171
Polymorphism, 72
Polystyrene, 39
Polytetrafluoroethylene, 171
Polyvinyl chloride, 38, 171
Post-heating, 290
Power spinning, 267
Precipitation, 84
   hardening, 66, 84, 147
Preheating, 290
Proeutectoid, 130
Proof strength, 201
Proportional limit, 200
Proton, 25
PVC, 171
Pyramid hardness, 184

Recrystallization welding, 290
Reduction of area, 202
Relief angles, 317
Resistance welding, 290
Rockwell hardness, 178
   indenter, 179
Rolling deformation, 261

Sachs, G., 262
SAE steel, 350
Saturated molecules, 35
Scholtky defect, 56
Scleroscope hardness, 186
Screw dislocation, 56
Self-diffusion, 114
Severity of quench, 140, 161
   H values, 163
Shaper, 312
Shear deformation, 267
   analysis, 270
   elastic, 270
   maximum strain, 271
   plastic, 271
Shear spinning, 267
Shear strain, 267
Shigley, J., 199
Shore scleroscope, 186
Side-cutting angle, 319
Side-rake angle, 318
Silver solder, 88
Simple hexagonal structure, 49
Single phase alloy, 65
Sintered carbide, 314
Slip, 31, 59
Slip lines, 62
Solid solution, 95
Solvent matrix, 55
Space lattice, 42
Specific volume, 109
Spheroidized pearlite, 99
Spot welding, 290
Stacking fault, 59
Static hardness, 187
Steel, 155
   alloy, 351
   carbon, 351
   cost effective, 155
   cost of, 166, 168
   designation of, 350
   free machining, 351

Index **375**

[Steel]
   high manganese, 351
   hypereutectoid, 102
   hypoeutectoid, 102
   stainless, 341
   transformation, 106
Strain, 188, 191
   analysis, 261
   compressive, 193
   cycle, 256
   deformation, 260
   equivalent, 256
   hardening, 182
   history, 263
   measurement, 260
   Meyer, 182
   shear, 267
   tensile, 193
Strain-hardening exponent, 182
Strain-strengthening rules, 253
Strength, 187
   analysis, 2, 250
   designation, 250
   direction, 250
   orientation, 250
   sense, 250
   type, 250
   of weldments, 287
Strength coefficient, 198
Strengthening exponent, 197, 204
Strengthening mechanisms, 63
   cold work, 64
   grain size, 63
   multiple phase, 65
   single phase, 65
Strength and microstructure, 297
Strength-nominal strain, 188
Stress, 188
Stress analysis, 6
Stress coefficient, 204
Stress concentration, 296
Stress-relieved, 158

Substitutional alloy, 77
Substitutional defect, 55
Substitution atom, 54
Surface diffusion, 116

Taylor equation, 309
Teflon, 171
Tempering, 141, 156
   curves, 142
   effect on hardness, 142
   effect on strength, 142
   ferrite, 142
   martensite, 142
   pearlite, 142
   source of data, 144
Tempering equations, 143
Tensile properties, 199
   area ratio, 203
   breaking strength, 202
   elastic limit, 200
   fracture strain, 203
   fracture strength, 203
   lower yield point, 202
   modulus of elasticity, 200
   proof strength, 201
   proportional limit, 200
   reduction of area, 202
   strain-strengthening exponent, 204
   stress coefficient, 204
   tensile strength, 202
   ultimate load strain, 202
   ultimate strength, 202
   uniform elongation, 205
   upper yield point, 202
   yield point, 201
Tensile property relationships, 204
   nominal strain to area ratio, 205
   strain strengthening exponent to cold work, 218
   tensile strength to Brinell hardness, 214
   tensile strength to cold work, 211

376 Index

[Tensile property relationships]
   tensile strength to fatigue strength, 239
   and strengthening exponent, 209
   true stress to cold work, 218
   true strain to nominal strain, 206
   true stress to nominal stress, 207
   ultimate load strain to strain-strengthening exponent, 207
   yield strength to cold work, 209
Tensile strength, 202
Ternary phase diagram, 69
Thermal conductivity, 323
Thermal strengthening, 155
Thermit welding, 290
Thermoplastic polymers, 40
Thermosetting polymers, 40
Tie line, 79, 80
Tool constants, 331
   cutting edge, 332
   effective rake angle, 332
   life, 331
   material, 331
Tool life, 311
Tool shape, 316
   back-rake, 317
   efficiency, 316
   end-cutting edge, 317
   nose radius, 320
   relief angles, 317
   side-cutting, 319
   side-rake, 318
Torsion deformation, 267
   analysis, 268
   maximum strain, 268
Transformation rate, 114
Transformation temperature, 106
True strain, 191
True stress, 188, 191
True stress-strain, 191
Twin boundary, 58

Ulastic deformation, 210
Ulastic strain, 210

Ultimate load, 202
Ultimate load strain, 202
Ultimate strength, 202
Unary phase diagram, 69
Undercooling, 108
Under yield, 199
Uniaxial deformation, 261
Uniform elongation, 205
Unit cell, 43
Unit elongation, 189
Unsaturated molecules, 36
Upper critical temperature, 101
Upper yield point, 202

Vacancy, 54
Valence, 26
Vanadium, 343
van der Waals bonds, 33
Vickers hardness, 184
Vinyl chloride, 38
Volume diffusion, 115

Water quench, 140
Wear resistance, 87
Welding processes, 288
   mechanical, 289
   mechanical and thermal, 288
   thermal, 290
   chemical and thermal, 292
Weld nugget, 292
Weld size, 292
   bead, 292
   calculation of, 292
   HAZ, 295
   penetration, 292
Whitney, E., 308
Work of deformation, 273

Yield load, 201
Yield point, 201
Yield strength, 200
Young's modulus, 60, 188, 200